2012
Nostradamus
Dragon Forecast
For All Signs

Dr. Louis Turi

2012 Nostradamus Dragon Forecast For All Signs
Dr. Louis Turi

© 2012 by Dr. Louis Turi
All rights reserved.

ISBN 978-1-257-00430-0

All rights reserved. No part of this publication may be reproduced or transmitted in any form or by any means, electronic or mechanical, including photocopy, recording, or any information storage and retrieval system without permission in writing from the publisher.

Startheme Publications Inc.
4411 N. 23rd Street
Phoenix, Arizona 85016

www.drturi.com

Printed and bound in the United States of America

Welcome to Each Sign of the Zodiac

Your Personal Horoscope for 2012

<u>Important Note from Dr. Turi:</u> I was born and raised in Provence, France. I rekindled and exercise only Nostradamus' 16th-century Divine Astrology method. This formula does not reflect the modern astrology disciplines you may use, study or practice. Realize that over 500 years ago the famous Prophet did not use a watch or any sophisticated computers. Thus like the great Seer, I investigate outer space and the Universal mind with my inborn spiritual telescope. A "microscopic attitude" will not help science or anyone else gain the Golden key to spiritual wisdom or Cosmic Consciousness. This limited expletive attitude is for scientists and astronomers alike who have long lost their spiritual values and replaced it with stationary scientifically oriented minds. We have all heard of the adage *"you can't see the forest for the trees."* Every one of them are aware of the twelve constellations of the Zodiac; but somehow it is still impossible for them to pass the limitation of their five rational senses and enter the intuitional domain of the stars. To penetrate the clear-sighted domain of those stars is a serious task that demands curiosity and a inborn advanced UCI (*Unique Celestial Identity*). But entering the archetypal realm of consciousness and

decoding the subtle meaning behind the symbols of the Zodiac within the structure of the Universal Mind involves more than a logical mind. You may learn much more about any and all subjects under the stars from my archived newsletters posted at http://www.drturi.com/newsletters/.

At this time, in space some scientists or religious souls are simply not allowed to see past their human senses during the course of this current incarnation. They do have eternity to bring forth their own cosmic consciousness in another incarnation as human are immortal spirits operating on a dense physical world called Earth. Realize that Divine Astrology *Astropsychology, is an extremely old celestial art and a very complex science and must be practiced as such. Not everyone is blessed with the "gift" needed to assimilate, understand and translate correctly the Divine order of the Creator and read the "signs." That's why a section of the bible clearly mentions, *"I will talk to you; but you won't hear me! I will present myself to you; but you won't see me!"* God speaks to us within his own celestial creation and manifestation and through his Divine spirit; only advanced souls will perceive his will. But the will (*or the part of God in each human being*) is stronger than the fears or skepticism and with education humanity can expand its own cosmic consciousness enabling the human masses with time, to "perceive and receive" the real manifestation of a Universal God. For those born on the cusp of any zodiacal sign, simply refer to the month of your birth, which reflects the exact constellation of your nativity. If you know your rising sign read your forecast for your ascendant too. Divine Astrology, as practiced by Nostradamus is the creation of Dr. Turi's *Astropsychology and its unique software that expresses a more accurate and simplistic way at handling and looking at the stars. All students have found it to be incredibly accurate. You may join Dr. Turi's Star student's family by taking the Astropsychology course by mail or in person. See http://www.drturi.com/lectures-and-courses/ for information.

Table of Contents

Capricorn 5
Saturn Governs the Power-Oriented, Structural Constellation of Capricorn

Aquarius 11
Uranus Governs The Ingenious, Freedom-Oriented Constellation of Aquarius

Pisces 17
Neptune Governs the Soft, Dreamy, Intuitive and Artistic Constellation of Pisces

Aries 23
Mars Governs the Aggressive - Warlike and Impatient Constellation of Aries

Taurus 29
Venus Governs the Beautiful and Financially Oriented Constellation of Taurus

Gemini 35
Mercury Governs the Nervous and Witty Dual Constellation of Gemini

Cancer 41
The Moon Governs the Nurturing and Caring Constellation of Cancer

Leo 47
The Sun Governs the Flamboyant and Majestic Constellation of Leo

Virgo 53
Mercury Governs the Precise and Critical Constellation of Virgo

Libra 59
Venus Governs the Diplomatic and Peaceful Constellation of Libra

Scorpio 65
Pluto Governs the Mighty Constellation of Scorpio "The Eagle or Lizard."

Sagittarius 71
Jupiter Governs the Philosophical and Educated Constellation of Sagittarius

Capricorn
Philosophical Astro – Poetry

Saturn Governs the Power-Oriented, Structural Constellation of Capricorn

Builder of the greatest towers
Holding all the social powers
Striving to climb to the highest peak
For honor has no place for the weak
I am Capricorn, child of Saturn.

Characteristics For Those Born In January

Saturn rules the practical sign of Capricorn and controls the month of January. You are strongly motivated to succeed and with dedication you will gain a position of power and respect in life. You are gifted with computers and you possess strong morals and organizational principles. More than any other sign of the Zodiac you strive for respect and accomplishments. Saturn is a karmic planet and rules your life, thus you must avoid nurturing depressing thoughts. The part of God in you is much stronger than the stars you inherited and you do have the power to master and use the Cosmic Code at your advantage. You were born in the middle of the winter when nature was asleep. You must have realized early on, that nothing would come easily to you. Like the goat, slowly but surely and against all odds (*cold / wind / snow*) you must climb towards the top of the mountain. The first part of your life will be a long and painful struggle but Saturn will reward you by giving you a long life and a well-deserved position at the end. You will appreciate old age and solid financial security. You may also marry a much younger or older partner. The fluctuations of the Moon strongly affect your mood and career success. The wise Capricorn soul will use his fish tail accordingly and synchronize his life and business with the Universal tides. Steadiness, organization, patience, and charm belong to you. You have a strong architectural or mathematical ability and your keen sense of observation will help you succeed in life. Karmic Saturn will exact payment for manipulating others for selfish ends and will throw the soul back to a painful start. You are attracted to power and successful people and you may marry into wealth. Emotional and sensitive, you are very responsible and protective of the family circle. However, you must learn to openly communicate your deep feelings.

Your real gifts are in your mind: Psychology, Astropsychology, interior designing, engineering, electronics, movie producer and any career related to Uncle Sam. Your own natal Dragon can propel you to the highest position and supreme power if you use it accordingly. But your challenge is to open up to the intangible world of the spirit and it's accompanying Universal rules. Your natural tendency to organize people and business at all times could hinder your sensitivity to others. Capricorns are good homemakers and adept with investments. As a rule you favor a successful business environment where you can apply your tremendous organizational gifts. **A word of caution for Capricorns:** Be aware of those wild acquaintances willing to help you to climb the ladder of success or the traps of religious poisoning. Remember to respect the Universal Law (*see Moon Power*), as your awareness of Moon planning will become a major contribution for happiness and success. The location of your natal Dragon's Head or Tail will seriously alter the strength or weakness of Saturn in your chart. The downfall of your spirit is religion and / or chemical addictions. Rush Limbaugh, Dr. Laura and Mel Gibson are good examples involving religion and / or chemical substance abuse. You can learn much more about yourself or anyone else by ordering any or all of my books entitled, *"I Know All About You," "The Power Of The Dragon", "And God Created The Stars"* or *"Beyond The Secret."*

2012 — Dragon Forecast For Those Born In January

Personal: On March 4th 2011, the karmic Dragon moved in the Axis Sagittarius *Head / Gemini *Tail and will stay there until August 30th 2012, stimulating your 12th and 6th houses regulating your subconscious creative / destructive forces and your work and service areas. Then in August 30th 2012, the Dragon changes its axis from Sagittarius / Gemini into Scorpio / Taurus and will stimulate all affairs regulating your wishes, friends, groups and your love, children, romance and speculations area.

On your 12th house (SUBCONSCIOUS) the Dragon's Head *positive will force you to undergo serious psychological changes making you very vulnerable to any induced fears based upon religious material. This impact will force you to "rebuild" your spirit appropriately with more progressive spiritual material or suffer the consequences of uncontrolled imagination that could lead to serious depressions and even suicidal tendencies. The Dragon's Tail in Gemini will also force the rebuilding of your atomic structure and will lead you to exercise or undergo worthwhile dieting programs and present a new you and a new image to the world. Indeed a very important year for you as you must be "reborn" in so many ways. The Dragon's Head will induce a great opportunity to establish a new spiritual fresh way at looking at the world and how you could participate helping others. The Dragon's Tail on your body will also help you to re-create a new image to the world while the unlucky Capricorn may suffer serious respiratory problems. The famous physicist Stephen Hawking, was born in January, and may suffer this Dragon drastically and find himself very close to God and the stars within the next two years. The months of June and December 2011 will be particularly trying for *All* of those born in January. Right on your health and work the Dragon's Tail will induce serious personal, physical changes and will not accept anything that might hurt your physical and mental health. The Dragon is demanding of you to learn a new spiritual way to perceive the world and may lead you to take on new studies. Be ready, and accept those changes with confidence and let go of the old fearful self.

Until August 30th 2012, matters involving the spirit, religions, health, and education will become powerful driving forces. The lucky souls born in January will enjoy a spiritual uplifting and a great dose of well controlled creative imagination. Many hard-working souls born in January will get the opportunity to improve their physical and spiritual images leading them to initiate good business decisions. Following a higher study many Capricorn souls will be promoted to a new higher position to service the world. This will open exciting doors to the world where traveling will also lead to more worthwhile business partnerships. The Dragon's Tail location in the sign of Gemini in your 6th house (WORK AND HEALTH) could bring havoc if you allow fears of the future to blind your spirit. Use your will and avoid disheartening thoughts when dealing with all the stress and fears the Tail of the Dragon may challenge you with. The disturbing Tail of the Dragon in health and work area will induce serious challenges and will force you to reevaluate your accepted spiritual convictions and realize the importance of neutralizing and eliminating any and all form of poisoning fears. Many unlucky Capricorn souls may find spiritual relief with legal / illegal drugs or alcohol making the situation much worse. More than any other signs of the zodiac, in 2011, souls born in January will need a serious dose of constant spiritual regeneration to counteract the relentless Dragon, if you feel its depressive, panicking impact, you MUST contact me and let me work on your spirit. One hour on the telephone or on Skype will do the job and bring back hope, health, faith and magic into your life. (602) 265-7667

2012 Predictions:

On August 30th 2012, the Dragon changes axis from Sagittarius / Gemini into Scorpio / Taurus.

The new Dragon's Head (LUCK / GROWTH) will affect your 11th house of wishes, groups and friends where you will experience serious progression and many of your wishes coming true. This impact may affect your sense of security and bring stress in the affairs of love, romance and children. The new dragon energy may stir a sense of freedom where the drive for adventures, speculations and responsibilities will conflict. Souls born in January will be tested to the limits in the affairs involving security, money, love and children, indeed 2012 could mark the end of important relationships that will directly affect their children and the support to them. The lucky Capricorn will be able to manage both and be greatly rewarded joining others to reach emotional, financial and spiritual stability. Many souls born in January will be starting or joining groups or associations where love, support and great opportunities will flourish.

JUPITER LUCKY TOUCH – In 2012, the great beneficial planet Jupiter (LUCK / EXPANSION / PROTECTION / TRAVELING / FOREIGNERS / STUDYING) will be cruising though the sign of Taurus (MONEY) until June 12th 2012. Jupiter's luck will benefit you in the affairs involving speculation, love and romance seriously altering the nasty effect produced by the Tail of the Dragon (KARMA) in this house. With Jupiter's blessings on your 5th house of creativity, love romance, children and speculations, expect gathering of rare wisdom from foreigners. The actions will bring the option to invest and make a lot of money (*or wisdom*) in the process. With Jupiter's blessings in Taurus affecting all financial institutions (BANKS) it could bring help from foreign countries. This impact could also be negative as foreign countries are buying / owing more and more real estate, offering the option for them to invest and make a lot of money in the process. When Jupiter (LUCK) moves in Gemini on June 12th 2012, expect more traveling and a wave of new banks restructures and information coming your way. You will enjoy foreign people, foreign ground and foreign food packing up extra pounds (*or serious wisdom*) in the process. Souls born in January should be ready for new associations and great business deals with those born in November, May, September and July. Working in good knowledge of your "Personal Lucky Dragon Window Dates" will also become a serious contribution to saving you time and money especially if you do a lot of traveling. Being at the right place at the right time has a lot to do with your progress in terms of lucky breaks and opportunities. In 2012, many souls born in January will be forced to reevaluate their work, health and service to the world. Many Capricorn will be offered (*or will chose*) the option to relocate in the process; The knowledge found in Astro-Carto-Graphy would be a major contribution to your success (*or your failure*) in one of these new locations. Keep this opportunity in mind and give it a try — it works! http://www.drturi.com/readings/ - Good luck to all souls born in January.

Aquarius
Philosophical Astro – Poetry

Uranus Governs The Ingenious, Freedom-Oriented Constellation of Aquarius

Holder of knowledge of the dimensions
The spark of all the inventions
Lover of all things in simplicity
Charged with the power of electricity
I am Aquarius, child of Uranus.

Characteristics For Those Born In February

The planet Uranus rules the sign of Aquarius and governs the month of February. You are one of the most original people walking this earth. Aquarius has produced many eccentric people and great inventors. Uranus rules the future and the incredible UFO phenomenon, technology, television, aeronautics, the Internet and humanitarianism. It commands NASA, Astronomy and all celestial knowledge particularly the old science of astrology and the Cosmic Code. You are blessed with curious stars and you are attracted to psychology, the food industry, real estate the police force and Astropsychology to name a few. Aquarius rules aeronautics, avionics, television, the Internet and advanced computers. The option to reach fame and fortune is a high probability during the course of your life if you service the world in an advanced and original way. The motion pictures "Back to the Future and The Matrix" are some of the best ways to illustrate Uranus' ingenuity in terms of artistic creativity. Strong and fixed, you have inherited from the stars, accurate intuition, tremendous common sense, ingeniousness, and a powerful will. Yet, you must learn to listen to others and participate in conversations with equality. Even when the ideas being presented are not of your own making, much knowledge can still be learned. Lend your full ear and do not race ahead with only thoughts of what you need to say.

Those born in February must also learn to positively direct Uranus' innovative mental power for the improvement and well being of the world. Acting eccentric and without forethought is a sure downfall for you. Your idealistic views are legendary and your mission is to promote Universal knowledge and Universal Brotherhood. You will benefit from the opportunity to use the latest technological arsenals to fulfill your unselfish wishes for mankind. You can handle the difficulties of life with a smile and transcend setbacks by using celestial knowledge to your benefit. The women of this sign are original, independent, beautiful, intellectual, and make good use of their incredible magnetic sexuality to reach their purposes. As a rule, women born in February produce extraordinarily intelligent children or twins. You are strongly advised not to eat when upset. The medical aspect of Divine Astrology predisposes those born in February to over-sensitive stomachs and an overactive mind. A word of caution for those born in February: Many young religious or rational souls will not understand your genius and your advanced message to the world. Many will try hard to stop and hurt you. Remember to respect the Universal Law (*see Moon Power*), as your awareness and Moon planning will become a major contribution to your happiness and success. The location of your natal Dragons Head or Tail will seriously alter the strength or weakness of Uranus in your chart. You can learn much more about yourself or anyone else by ordering my new book entitled *"I Know All About You," "The Power Of The Dragon" "Beyond The Secret" "And God Created The Stars."*

2012 — Dragon Forecast For Those Born In February

Personal: On March 4th 2011, the karmic Dragon moved in the Axis Sagittarius *Head / Gemini *Tail and will stay there until August 30th 2012, stimulating your 5th and 11th houses regulating love, romance, children, speculations and your 11th house of groups, wishes and friends. Then in August 30th 2012, the Dragon changes its axis from Sagittarius / Gemini into Scorpio / Taurus and will stimulate all affairs regulating your public standing and your home life area.

In your 11th house (GROUPS / FRIENDS) the Dragon's Head *positive will offer you great opportunities to reach a bigger and higher place in the world making you very much in demand all based upon your endless drive to educate the world. This impact will force you to "rebuild" a new, more diplomatic you with great progressive material to share with others; this could lead you to more recognition even fame. The Dragon's Tail in Gemini keeps bringing mental stress being in demand by all. This Dragon may also induce stress with loved ones or children wanting more of your attention. Remember your priorities and return love to those who need it most. Indeed a very important year for you as you are forced to expand to fulfill your mission and reach and teach from all corners of the world. The current Dragon's Head in Sagittarius will induce great contacts from foreigners or from foreign grounds, with it the opportunity to introduce a new spiritual fresh way at looking at the world. Many opportunities will be offered to you on how you could participate helping others. The Dragon's Tail in your creativity wants you to adjust and adapt while being productive all along as to re-create a new image to the world. While the unlucky Aquarius may suffer relationships challenges, patience and understanding is the key. The month of June 2012 will be particularly trying for all of those born in February while the lucky one will benefit from good karma. Right "on your love, romance, children and creativity house the Dragon's Tail will induce serious personal challenges. The Dragon is demanding of you to join others for strength without hurting those you care. Be ready to accept those changes with confidence be productive with all.

Until August 30th 2012, matters involving group association, traveling, publishing and foreigners will become powerful driving forces. Deserving souls born in February will enjoy new friends promoting many of their wishes. Many hard-working Aquarius souls will get plenty of opportunities to expand, learn and teach following good business decisions. Some traveling will also offer you with new and valuable promotions and a better higher position to reach the world. The Dragon will open exciting new doors to reach the world that will also lead many of your friends to more worthwhile business partnerships. The Dragon's Tail location in the sign of Gemini in your 5th house (SPECULATION) could bring havoc if you rush too fast. Use your will and avoid negative thoughts when dealing with all the stress and fears the Tail of the Dragon may challenge you with. The disturbing Tail of the Dragon does not have to win over you but will induce serious challenges with loved ones forcing you behavior or need for independence. Many unlucky Aquarius souls may find freedom at the expenses of those they care just to realize their error in the future. Souls born in February will need to think before acting and hurt those they care for. If you feel its depressive, panicking impact you MUST contact me and let me work on your spirit. One hour on the telephone or on Skype will do the job and bring back hope, health, faith and magic into your life. (602) 265-7667

2012 Predictions:
On August 30th 2012, the Dragon changes axis from Sagittarius / Gemini into Scorpio / Taurus.

The new Dragon's Head (LUCK / GROWTH) will stimulate your 10th house of career where you will experience a serious expansion where many of your career wishes will come true. This impact will upgrade your sense of security and may force you to move or expand the base of your operations. The new dragon energy will also stir *Mr. Ego* and *Mrs. Pride* with a new drive for adventures, something you must keep in check. Souls born in February will be tested to the limits in the affairs involving home and security but 2012 could mark the beginning of an era that will directly affect their position in the world. The lucky Aquarius soul will be able to manage both a very busy career and home responsibility leading to emotional, financial and spiritual stability. Many deserving souls born in February will see the results of many years of arduous work where more opportunities will flourish.

JUPITER LUCKY TOUCH – In 2012, the great beneficial planet Jupiter (LUCK / EXPANSION / PROTECTION / TRAVELING / FOREIGNERS / STUDYING) will be cruising though the sign of Taurus (MONEY) until June 12th 2012. Jupiter's luck will benefit you in the affairs involving your base of operation, real estate, family matters seriously altering the nasty effect produced by the Tail of the Dragon (KARMA) in this house. With Jupiter's blessings in Taurus affecting all financial institutions (BANKS) it could bring help from foreign countries. This impact could also be negative as foreign countries are buying / owing more and more prime real estate, offering the option for them to invest and make a lot of money in the process. When Jupiter (LUCK) moves in Gemini on June 12th 2012, expect more traveling and a wave of information pertaining to banks restructures coming your way. You will enjoy your own opportunities to invest on foreign ground in the process. Souls born in February should be ready for great career associations and great business deals with souls born in June, August, October and those born in December. Working in good knowledge of your "Personal Lucky Dragon Window Dates" will also become a serious contribution to saving you time and money especially if you do a lot of traveling. Being at the right place at the right time has a lot to do with your progress in terms of lucky breaks and opportunities. In 2012, many souls born in February will be forced to reevaluate their career and security. Many Aquarius souls will also be offered (*or will chose*) the option to relocate in the process; The knowledge found in Astro-Carto-Graphy would be a major contribution to your success (*or your failure*) in one of these new locations. Keep this opportunity in mind and give it a try — it works! http://www.drturi.com/readings/ - Good luck to all souls born in February.

Pisces
Philosophical Astro – Poetry

Neptune Governs the Soft, Dreamy, Intuitive and Artistic Constellation of Pisces

Mystical and magical
Nebulous and changeable
I work my way up life's rivers and seas
To my place at God's own feet
I am Pisces, child of Neptune.

Characteristics For Those Born In March

The planet Neptune and the sign of Pisces govern the month of March. You are a natural teacher, a philosopher and a perfectionist of the soul. You inherited a phenomenal intuition and you will exercise more intuition than logic in dealing with life in general. You are a gifted artist and you enjoy holistic endeavors, your love for animals is unsurpassed and you will always do your best to love and protect them. Many advanced Pisces are also involved in the medical profession, writing and teaching. The young Pisces soul may also work in the construction fields. However Pisces must understand the importance of higher education if he is to use his full potential and teaching gifts. You are noted for your sensitivity, creativity, and artistic values. Michelangelo, Einstein and George Washington were also Pisces' and used their creativity to the fullest. Your downfall is an over preoccupation with others, guilt feelings, addictions and a blind acceptance of religious dogmas, (*i.e., Pastor Joel Osteen*). Nevertheless, your good heart is not surpassed by any other sign of the zodiac and the advanced Pisces possess spiritual healing powers and true Universal wisdom. Highly evolved people born in March will lead many lost souls out of the deep clouds of deception towards the true colors of love and cosmic consciousness.

Your soul's purpose is to swim upstream towards the ethereal light of oneness to find God. A young March spirit is deceiving, complaining and addicted to religious dogmas, cult endeavors, chemicals, drugs, and alcohol. Pisces is a karmic sign and has within itself the potential to reach immortality, fame and fortune through artistic or spiritual work. In the medical aspect of Divine Astrology, Pisces rules the feet. It is important for you to walk barefoot on the grass to regenerate the body through the magnetic fields of the earth. Your intuition is remarkable and should be well heeded when confronted with serious decisions. A word of caution for Pisces: Do not swim downstream as your induced faith could take you to Neptune's deepest quicksand with no option for return. David Koresh, Rev. Jim Jones and Harold Camping are good examples of Neptune's deceiving religious Captains. As a water sign, remember to respect the Universal Law (*see Moon Power*), your awareness and respect of the Moon's fluctuations will become a major contribution to your happiness and success. The location of your natal Dragons Head or Tail will seriously alter the strengths or weakness of Neptune in your chart. You can learn much more about yourself or anyone else by ordering my new book entitled, *"Beyond the Secret," "I Know All About You," "The Power Of The Dragon"* or *"And God Created The Stars."*

2012 — Dragon Forecast For Those Born In March

Personal: On March 4th 2011, the karmic Dragon moved in the Axis Sagittarius *Head / Gemini *Tail and will stay there until August 30th 2012, stimulating your 10th and 4th houses regulating general security and career matters. Then in August 30th 2012, the Dragon changes its axis from Sagittarius / Gemini into Scorpio / Taurus and will stimulate all affairs regulating your mind, higher learning and publishing.

In your 4th house (HOME / REAL ESTATE) the challenging Dragon's Tail will impose a restructure of your basic security and the option to upgrade or relocate close or away from the water to enjoy calmness and the wild. There you will be able to channel more from the spirits fulfilling your endless creativity and the drive to educate others. These changes will force you to "rebuild" a new spirit and reach the world electronically. Your creative mind could also lead you to more recognition even fame if you associate with foreigners or aim for foreign grounds. The Dragon's Tail in Gemini keeps bringing obstacles and frustration related to security, home and family altering your actions and wisdom, patience is the key. This Dragon may also induce stress with siblings and transportation (*i.e.*, UK *deadly chain reaction*). Remember, the Dragon's priority is to help you reevaluate your wisdom, beliefs and upgrade to a higher perception. Indeed 2011 is an important year forcing you to expand mentally and fulfill your teaching mission. The current Dragon's Head in Sagittarius will bring about great association or even contacts from foreigners dedicated to help you as you help them. With it the opportunity to introduce a new spiritual fresh way at looking at education in the world. Many opportunities will be offered to you as you upgrade your status and your career and how you could also participate in helping the world. The Dragon's Head in your career area wants you to adjust and adapt while being productive all along bringing your gifts to the world. While the unlucky Pisces may suffer a total home collapse, taking on the challenges of the current economy with patience will make you more efficient in reaching your wishes. The month of June 2012 will be particularly trying for all of those born in March while the lucky Pisces will benefit from good karma. You may also be forced to consider relocation then. Right on your public standing house the blessings of the Dragon's Head Tail will induce wonderful opportunities. The Dragon is demanding of you to adjust to a new environment, to stimulate your career and join others for strength. Be ready to accept those changes with confidence the future has a lot to offer you.

Until August 30th 2012, matters involving career and group association, traveling, publishing and foreigners will become powerful driving forces for Pisces. Deserving souls born in March will enjoy a new base of operation and new friends promoting many of their wishes. Many hard-working Pisces souls will get plenty of opportunities to expand, learn and teach following good business decisions. Some traveling will offer you with new and valuable opportunities and with time, a better higher position in the world. The Dragon will open exciting new doors leading many of your friends to become involved in business partnerships. The Dragon's Tail location in the sign of Gemini in your security house could bring havoc if you rush or try to hard. Use your will and avoid negative thoughts when dealing with all the stress and fears the Tail of the Dragon may challenge you with. The disturbing Tail of the Dragon does not have to win over you but will induce serious restructuring challenges where patience becomes the key. Many Pisces souls may find themselves rebuilding their entire life and this could stir fears of the future. Souls born in March will need to adjust to a new life and if you feel the Dragon's depressive, panicking impact you MUST contact me and let me work on your spirit. One hour on the telephone or on Skype will do the job and bring back hope, health, faith and magic into your life. (602) 265-7667

2012 Predictions:

On August 30th 2012, the Dragon changes axis from Sagittarius / Gemini into Scorpio / Taurus.

The new Dragon's Head (LUCK / GROWTH) will stimulate your 9th house (HIGHER LEARNING'S / UNIVERSITIES / COLLEGES) where you will experience a rebirth of your own consciousness helping others to get to your own spiritual level. The Dragon will induce tremendous intellect discerning power and an enormous surge of creativity. This impact will upgrade your spiritual senses and induce channeling and a urge for writing. The new dragon energy will also demand of you to constantly regenerate with creative or investigative material. Souls born in March will be tested to the limits in the affairs involving the mind, learning, teaching, publishing inducing traveling. The year 2012 could mark the beginning of an era that could lead them to a teaching well respected position. The lucky Pisces soul will be able to manage both the physical and spiritual realms promoting them towards a very busy career. Many deserving souls born in March will begin to see the results of many years of arduous mental work where more opportunities will flourish.

JUPITER LUCKY TOUCH – In 2012, the great beneficial planet Jupiter (*LUCK / EXPANSION / PROTECTION / TRAVELING / FOREIGNERS / STUDYING*) will be cruising though the sign of Taurus (*MONEY*) until June 12th 2012. Jupiter's luck will add benefit to the 3rd house affairs related to intellectual properties, teachings and publishing. The challenging effect produced by the Tail of the Dragon (*KARMA*) in Taurus will be seriously altered with Jupiter's protection. With Jupiter's blessings in Taurus affecting all financial institutions (*BANKS*) it could bring help from foreign countries. This impact could also be perceived as negative as foreign countries are buying / owing more and more prime real estate, offering them the option to invest and make a lot of money in the process. When Jupiter (*LUCK*) moves in Gemini on June 12th 2012, expect an upsurge of information pertaining to banks restructures coming your way. You could also benefit and invest on foreign ground in the process. Souls born in March should be ready for mental exploration and beneficial career associations offering great prospective business deals with souls born in September, January, July and those born in November. Working in good knowledge of your "Personal Lucky Dragon Window Dates" will also become a serious contribution to save you time especially if you do a lot of traveling. Being at the right place at the right time has a lot to do with your progress in terms of lucky breaks and opportunities. In 2012, many Pisces will be forced to reevaluate their spiritual values where learning and teaching will become the main focus. Many Pisces will also be offered (*or will chose*) the option to relocate and reach a much bigger world in the process; The knowledge found in Astro-Carto-Graphy would be a major contribution to your success (*or your failure*) in one of these new locations. Keep this opportunity in mind and give it a try — it works! http://www.drturi.com/readings/ - Good luck to all souls born in March.

Aries

Philosophical Astro – Poetry

Mars Governs the Aggressive - Warlike and Impatient Constellation of Aries

All will hear my views and voice
Trial and error is my school of choice
Like a dragon, dashing and daring I appear
Fighting for those that I hold dear
I am Aries, child of Mars.

Characteristics For Those Born In April

Assertive Mars controls the month of April. In Greek Mythology, this planet is called "The Lord of War," and rules the impatient sign of Aries. You were born a leader however, because your inborn impatience you may also learn by making a few mistakes. Your strong and impatient desire to succeed must be controlled and hasty decisions avoided. Others perceive you as a competitive and motivated child like person. More than any other sign of the zodiac, souls born in April must learn steadiness, organization and most of all diplomacy. When confronted, grace and charm does not really belong to you. Martian souls possess strong leadership and engineering abilities and April men are attracted to dangerous sports, speed, engineering, and the military. Due to your "turbocharged" personality, you are also accident-prone to the head, and should protect it at all times. Both male and females born in April tend to talk too much and must learn to listen to others and control impatience. You must focus on your needs steadily and finish what you have started. Inadvertently the "red" uncontrolled Martian personality will hurt sensitive souls; thus damaging the chances for respect and promotion.

Your explosive temper is generated by an inborn fear of rejection and an inner inferiority complex. Do not take rejection or opposition personally. The "childlike" attitude could attract manipulative spirits wishing to structure or use the immense creativity and energy of the Mars competitive spirit. You do love your home and you are responsible with your family. Nevertheless, you prefer to be where the action is, as you get bored easily. If you practice patience, tolerance and diplomacy, there is no limit to where Mars will take you and this is all the way to the highest level of accomplishment. Your main lesson is to learn all the diplomatic and loving traits of the opposite Venus-ruled sign Libra. Some young April souls are totally consumed with themselves and will not share possessions or the light of the stage with others supporters. Once you find yourself and confidence, the option to become a leader of the mind in any chosen field will be given to you. Souls born in the month of April must assume a diplomatic attitude when dealing with others and when dealing with corporate money reward those who helped them. Word of caution; dealing with any Aries soul demands anyone to be very aware of a natural drive to control all corporate financial areas due to a subconscious drive and fear of power, (i.e., Bernie Madoff and Conspiracy champion author, David Icke... note both souls were born April 29 or 9 days after Adolf Hitler also born in April). The location of your natal Dragon's Head or Tail will seriously alter the strengths or weaknesses of Mars in your chart. You can learn much more about yourself or anyone else by ordering my new book entitled *"Beyond The Secret," "I Know All About You," "The Power Of The Dragon"* or *"And God Created The Stars."*

2012 — Dragon Forecast For Those Born In April

<u>Personal:</u> On March 4th 2011, the karmic Dragon moved in the Axis Sagittarius *Head / Gemini *Tail and will stay there until August 30th 2012, stimulating your 3rd and 9th houses regulating your mental power your higher learning potential and publishing. Then in August 30th 2012, the Dragon changes its axis from Sagittarius / Gemini into Scorpio / Taurus and will stimulate all affairs regulating your 8th house (CORPORATE ENDEAVORS) and your 2nd house (MONEY).

The challenging Dragon's Tail will impose a restructure of your critical mental power inducing an avalanche of endless thoughts leading to mental exhaustion. Channeling and writing will stimulate your endless creativity and the drive to educate others but you will need to slow down, stop and recharge your own mind. The mental challenges are simply exhausting as you want to reach more of the world electronically.

The difficult Dragon afflicting your mind brings about outside interferences from your past and induce uncontrolled imagination fuelled by fears of personal or career failures. The current progressive Dragon's Head in Sagittarius will enhance your 9th house (FOREIGNERS) and stimulate creativity leading you to more recognition even fame if you associate with foreigners or aim for foreign grounds. The Dragon's Tail in Gemini keeps bringing obstacles and frustration altering your mental clarity, actions and wisdom, patience is the key. This Gemini Dragon's Tail may also induce stress with siblings and danger with transportation. (*i.e., UK deadly chain reaction. At least 7 dead, 51 hurt in traffic 'fireball' in Britain.*)

Remember, the Dragon's priority is to help you reevaluate your wisdom, beliefs and upgrade to a higher perception of your mental self. Indeed 2011 is an important year forcing you to expand mentally and fulfill your teaching mission. The current Dragon's Head in Sagittarius will bring about great association or even contacts from foreigners dedicated to help you as you help them. With it the opportunity to introduce a new spiritual fresh way at looking at education in the world at large. Many opportunities will be offered to you as you upgrade your perception and your teaching, learning potential and how you could also participate in helping the world. The Dragon's Head in your traveling house and learning / teaching area wants you to adjust and adapt while being productive all along bringing your gifts to the world. While the unlucky Aries may suffer total exhaustion, taking on all the challenges with patience will make you more efficient in reaching your wishes and avoid depressions. The month of June 2012 will be particularly trying for all of those born in April while the lucky Aries will benefit from good karma. Right on your mental houses the blessings of the Dragon's Head Tail will induce wonderful opportunities to travel and teach and be even published. The Dragon is demanding of you to adjust to a new form of thinking, learning, teaching to stimulate your career and reach for a higher wisdom for mental strength. Be ready to accept those changes with confidence because the future has a lot to offer you.

Until August 30th 2012, matters involving traveling, learning, teaching publishing and foreigners will become powerful driving forces for Aries. Deserving souls born in April will benefit from foreign friends / teachers promoting many of their wishes. Many hard-working Aries souls will get plenty of opportunities to expand following good business decisions. Some traveling opportunities will offer you a new understanding and a higher position in the world. The Dragon's Tail location in the sign of Gemini afflicting your mind could bring havoc if you give in to an uncontrollable imagination, rush or try to too hard. Use your will and avoid negative thoughts when dealing with all the stress and fears the Tail of the Dragon may challenge you with. The disturbing Tail of the Dragon does not have to win over you but will induce serious mental challenges where patience becomes the key. Many Aries souls may find themselves rebuilding their entire life and this could stir fears of the future. Souls born in April will need to adjust to a way of thinking and if you feel the Dragon's depressive, panicking impact you MUST contact me and let me work on your spirit. One hour on the telephone or on Skype will do the job and bring back hope, health, faith and magic into your life. (602) 265-7667

2012 Predictions:
On August 30th 2012, the Dragon changes axis from Sagittarius / Gemini into Scorpio / Taurus.

The new Dragon's Head (LUCK / GROWTH) will stimulate your 8th house (CORPORATE MONEY / CONTRACTS / LEGACY) where you will experience a rebirth of your own aptitude to generate or make an income with others. The Dragon will induce a new discerning financial power and with it an enormous surge of ideas leading to investment and financial creativity. This impact will upgrade your sense of security working within the structure of powerful well established groups. The new dragon energy in your finances will also demand of you to constantly regenerate with creative or investigative material. Souls born in April will be tested to the limits in the affairs involving big money making schemes. Abusive April souls will also feel the impact of an overdue karma with Uncle Sam while others are strongly cautioned to stay clean of criminal elements. The year 2012 could mark the beginning of an era that could mean financial security or the loss of it all for many Aries. Others will enjoy a well deserved, well respected position. The lucky Aries soul will be able to manage both the physical and spiritual realms promoting them towards a successful year. Many deserving souls born in April will begin to see the results of many years of arduous mental work where more opportunities will flourish.

JUPITER LUCKY TOUCH – In 2012, the great beneficial planet Jupiter (LUCK / EXPANSION / PROTECTION / TRAVELING / FOREIGNERS / STUDYING) will be cruising though the sign of Taurus (MONEY) until June 12th 2012. Jupiter's luck will add benefit to the 2nd house affairs related to all money making schemes. The challenging effect produced by the Tail of the Dragon (KARMA) in Taurus will be seriously altered with Jupiter's protection. With Jupiter's blessings in Taurus affecting all financial institutions (BANKS) it could also bring help from foreign countries. This impact could also be perceived as negative as foreign countries are buying / owing more and more prime real estate, offering them the option to invest and make a lot of money in the process. When Jupiter (LUCK) moves in Gemini on June 12th 2012, expect an upsurge of information pertaining to banks restructures coming your way. You could also benefit and invest on foreign ground in the process. Souls born in April should be ready for beneficial career associations offering great prospective business deals with souls born in August, December, October and those born in February. Working in good knowledge of your "Personal Lucky Dragon Window Dates" will also become a serious contribution to saving you time especially if you do a lot of traveling. Being at the right place at the right time has a lot to do with your progress in terms of lucky breaks and opportunities. In 2012, many Aries will be forced to reevaluate the way they handle and will become the main focus. Many Aries will also be offered (*or will chose*) the option to relocate and reach a much bigger world in the process; The knowledge found in Astro-Carto-Graphy would be a major contribution to your success (*or your failure*) in one of these new locations. Keep this opportunity in mind and give it a try — it works! http://www.drturi.com/readings/ - Good luck to all souls born in April.

Taurus

Philosophical Astro – Poetry

Venus Governs the Beautiful and Financially Oriented Constellation of Taurus

Luxurious and elegant
I have the memory of an elephant
Loving all of life's finer pleasures
Gifted am I at acquiring more coffers and treasures
I am Taurus, child of Venus.

Characteristics For Those Born In May

The month of May is governed by the planet Venus and by the reliable sign of Taurus. Others perceive you as beautiful, somehow stubborn and practical. You are the money-maker sign of the Zodiac and you have stability and true love to offer to others. You need to control your jealousy, insecurity, and your authoritarian attitudes and avoid eating when upset. You are a gifted artist and strive for organization. You are also attracted to the professions of banking, real estate, the arts, computers, radio, television, Astropsychology, aeronautics food, real estate and investigation, to name a few. Many "Bulls" will reach fame and fortune and enjoy the security of a beautiful and big house. Strong and dominant, you have inherited a deep intuition, a tremendous common sense, and a powerful will. Venus rules love and possession; you must avoid destructive thoughts pertaining to jealousy, stubbornness and insecurity. Learn to channel Venus' constructive powers towards creativity, diplomacy and love. If you behave in an insecure stubborn sarcastic, destructive unattractive manner you will lose it all in the end. Your down-to-earth approach to life must not interfere with your spiritual growth.

Part of your lesson in this lifetime is to keep an open mind to the world of the spirit and use the metaphysical information to ensure financial growth. Your desire for practicality and riches is legendary but you will always regenerate with New Age and metaphysical matters. You will courageously handle the difficulties of life with a solid attitude, and you inherited a beautiful nobility of purpose. Girls born in May are beautiful, classic, intellectual, magnetic, and sensitive, and will always combine Venus' beauty and sensual magnetism to attain their goals. You are meticulous and critical about your mate and it is important for you to marry someone well groomed and well respected. With you, love must last forever. Food is often on your mind, again, do not eat when you are upset. Remember to respect the Universal Law (*see Moon Power*), as your awareness and moon planning will become a major contribution towards reaching many of your dreams. The location of your natal Dragons Head or Tail will seriously alter the strengths or weakness of Venus in your chart. You can learn much more about yourself or anyone else by ordering any of my books titled *"Beyond The Secret," "I Know All About You," "The Power Of The Dragon"* or *"And God Created The Stars."*

2012 — Dragon Forecast For Those Born In May

<u>Personal:</u> On March 4th 2011, the karmic Dragon moved in the Axis Sagittarius *Head / Gemini *Tail and will stay there until August 30th 2012, stimulating your 2nd and 8th houses regulating your personal and corporate income. Then in August 30th 2012, the Dragon changes its axis from Sagittarius / Gemini into Scorpio / Taurus and will stimulate all affairs regulating your 1st house (YOURSELF) and your 7th house (MARRIAGE / CONTRACTS).

The challenging Dragon's Tail will impose a restructure of your personal finances and self esteem inducing an endless wave of insecurity forcing to adapt to the current depressive economy. Investigating new ways out and writing will stimulate your endless creativity and the drive to reach for better solutions. You will need to stop worrying about others and recharge your own mind. The mental challenges are simply exhausting as you want to help those you care deeply. The difficult Dragon afflicting your self esteem brings about outside interferences from your past and induce uncontrolled imagination fuelled by fears of personal or career failures. The current progressive Dragon's Head in Sagittarius will enhance your 8th house (CORPORATE AFFAIRS / INVESTMENTS) and stimulate creativity leading to new financial opportunities if you associate with foreigners or aim for foreign grounds. The Dragon's Tail in Gemini keeps bringing obstacles and frustration altering your mental clarity, actions and wisdom; patience is the key. This Gemini Dragon's Tail may also induce stress with siblings and danger with transportation. (i.e., *UK deadly chain reaction. At least 7 dead, 51 hurt in traffic 'fireball' in Britain.*)

Remember, the Dragon's priority is to help you reevaluate your financial aptitudes and upgrade to a higher perception of your creative self. Indeed 2011 is an important year forcing you to expand mentally and fulfill your dream for endless financial security. The current Dragon's Head in Sagittarius will bring about great association or even contacts from foreigners dedicated to help you as you help them. With it the opportunity to introduce a new spiritual fresh way at looking at finances and education in the world at large. Many opportunities will be offered to you as you upgrade your perception of the divine and your learning potential and how you could also participate in helping the world to reach stability. The Dragon's Head in your investment house, learning / teaching area wants you to adjust and adapt while being productive all along bringing your own gifts to the world. While the unlucky Taurus may suffer total financial exhaustion, taking on all the challenges with patience will make you more efficient in reaching your wishes and avoid depressions. The month of June 2012 will be particularly trying for all of those born in May while the lucky Taurus will benefit from good karma. Right on your financial houses the blessings of the Dragon's Head Tail will induce wonderful opportunities to travel and teach and be even published. The

Dragon is demanding of you to adjust to a new form of thinking about possession and wealth while learning, teaching will stimulate your career. Be ready to accept those changes with confidence because the future has a lot to offer you.

Until August 30th 2012, matters involving finances, traveling, learning, teaching publishing and foreigners will become powerful driving forces for Taurus. Deserving souls born in May will benefit from foreign friends / teachers promoting many of their wishes. Many hard-working Tauruses will get plenty of opportunities to expand following good business decisions. Some traveling opportunities will offer you a new understanding and a higher position in the world. The Dragon's Tail location in the sign of Gemini afflicting your self esteem could bring havoc if you give in to an uncontrollable imagination, rush or try to too hard. Use your will and avoid negative thoughts when dealing with all the stress and fears the Tail of the Dragon may challenge you with. The disturbing Tail of the Dragon does not have to win over you but will induce serious mental challenges where patience becomes the key. Many Taurus souls may find themselves rebuilding their entire financial life and this could stir fears of the future. Souls born in May will need to adjust to a way of thinking and if you feel the Dragon's depressive, panicking impact you MUST contact me and let me work on your spirit. One hour on the telephone or on Skype will do the job and bring back hope, health, faith and magic into your life. (602) 265-7667

2012 Predictions:

On August 30th 2012, the Dragon changes axis from Sagittarius / Gemini into Scorpio / Taurus.

The new Dragon's Head (LUCK / GROWTH) will stimulate your 7th house (MARRIAGE / CONTRACTS / PARTNERSHIPS / LEGACY) where you will experience a rebirth of your own aptitude to generate or make an income with others. The Dragon will induce a new discerning financial power and with it an enormous surge of ideas leading to investment and financial creativity. This impact will upgrade your sense of security working within the structure of powerful well established groups or with worthwhile partners. The new dragon energy in your marriage area will open many doors especially with foreigners where love can bless your life. Souls born in May will be tested to the limits in the affairs involving marriages and karmic relationships that must come to an end. Abusive May souls will also feel the impact of an overdue karma with Uncle Sam while others are strongly cautioned to stay clean of the criminal elements. The year 2012 could mark the beginning of an era that could mean enjoying or ending a long lasting relationships. Others will enjoy a well deserved, well respected partner and a higher position in the world of finances. The lucky Taurus soul will be able to manage both their business and emotional partnerships offering them a successful year. Many deserving souls born in May will begin to see the results of many years of arduous mental work where more opportunities will flourish.

JUPITER LUCKY TOUCH – In 2012, the great beneficial planet Jupiter (LUCK / EXPANSION / PROTECTION / TRAVELING / FOREIGNERS / STUDYING) will be cruising though the sign of Taurus (MONEY) until June 12th 2012. Jupiter's luck will add benefit to the 7th house affairs related to partnerships and deals from foreign grounds. The challenging effect produced by the Tail of the Dragon (KARMA) in Taurus will be seriously altered with Jupiter's protection. With Jupiter's blessings in Taurus affecting all financial institutions (BANKS) it could also bring help from foreign countries. This impact could also be perceived as negative as foreign countries are buying / owing more and more prime real estate, offering them the option to invest and make a lot of money in the process. When Jupiter (LUCK) moves in Gemini on June 12th 2012, expect an upsurge of information pertaining to banks restructures coming your way. You could also benefit and invest on foreign ground in the process. Souls born in May should be ready for beneficial career associations offering great prospective business deals with souls born in November, September, January and those born in March. Working in good knowledge of your "Personal Lucky Dragon Window Dates" will also become a serious contribution to save you time especially if you do a lot of traveling. Being at the right place at the right time has a lot to do with your progress in terms of lucky breaks and opportunities. In 2012, many Tauruses will be forced to reevaluate the way they handle partnerships and finances making it the main focus. Taurus will also be offered (*or will chose*) the option to relocate and reach a much bigger world in the process; The knowledge found in Astro-Carto-Graphy would be a major contribution to your success (*or your failure*) in one of these new locations. Keep this opportunity in mind and give it a try — it works! http://www.drturi.com/readings/ - Good luck to all of those born in May.

Gemini

Philosophical Astro – Poetry

Mercury Governs the Nervous and Witty Dual Constellation of Gemini

Freethinking and intelligent
You will not find me under rigorous management
You may think you know me well
Then my other half over you casts a spell
I am Gemini, child of Mercury.

Characteristics For Those Born In June

The planet Mercury rules the sign of Gemini. You are intellectual, nervous and adaptable. Born with a strong drive to communicate, you are classified in Greek mythology as "The Messengers of the Gods." You inherited a gift of youth, a double personality and a quicksilver mind enabling you to adapt easily to any situation. On a negative note, Mercury, the "Lord of the Thieves" breeds volatile and unreliable people due to their dual characteristics. You are a gifted communicator and radio, language, photography, sales, movies, acting, dancing and the medical field and any type of public relations work appeals to you. Your natural speed for life's experiences makes you impatient and nervous. You must learn to focus and crystallize your powerful mind. You have the potential to become an efficient speaker and produce interesting books. Due to your strong desire for security, many of you will be attracted to the real estate and food industries. Your financial potential is unlimited if you learn and make a good use of the Universal Law in charge of your 2nd house of income. Strong Mercury will produce an incredible amount of physical and spiritual energy that must be dissipated. The unaware psychological fields classify those children as having ADD. (*"Attention Deficit Disorder"* or *ADHD*).

Contrary to what scientists assume and perceive as an indisposition, ADHD it is actually a potent gift from God. The soul is simply programmed to naturally reject traditional education, thus opening the rare door to genius and with it the potential for new discovery. Incidentally, President Clinton, Einstein, and Dr. Turi were born with an "ADHD affliction." Thus if a teacher is mistaken about some information, the Mercurial soul's inborn sense of curiosity and discovery will bring about potential information leading to the truth. Impatience, nervousness, mental curiosity, and a short attention span are your characteristics. You will never follow long established dogmas. Your Mercurial spirit will open new doors to mental exploration. You are curious by nature and are always questioning. Boredom is your worst enemy and you must associate with intellectual people who can stimulate your incredible mind. Telling jokes is also a part of your mental agility. A word of caution for you: Always be alert when the Moon crosses the deadly sign of Scorpio at work, especially after the Full Moon. Remember to respect the Universal Law as your awareness and moon planning will be a major contribution of avoiding dramatic experiences, and will help you reach many of your dreams. The location of your natal Dragon's Head or Tail will seriously alter the strength or weakness of Mercury in your chart. You can learn much more about yourself or anyone else by ordering my new book entitled *"Beyond The Secret," "I Know All About You," "The Power Of The Dragon"* or *"And God Created The Stars."*

2012 — Dragon Forecast For Those Born In June

Personal: On March 4th 2011, the karmic Dragon moved in the Axis Sagittarius *Head / Gemini *Tail and will stay there until August 30th 2012, stimulating your 1st and 7th houses regulating your self discovery and your partnership / contract / marriage house. Then in August 30th 2012, the Dragon changes its axis from Sagittarius / Gemini into Scorpio / Taurus and will stimulate all affairs regulating your 12th house (SUBCONSCIOUS) and your 6th house (SERVICE TO THE WORLD).

The challenging Dragon's Tail will impose a restructure of your physical self where you will be driven to look your best physically, this impact will induce an avalanche of endless thoughts to uncover who you are and how other people sees you. Your need for love and light will stimulate your channeling and writing with a drive to educate others but you will need to slow down, stop and recharge your own mind. The mental challenges are simply exhausting trying to bring the light to the world. The difficult Dragon afflicting your self brings about outside interferences from your past and induce uncontrolled imagination fuelled by fears of personal or career failures.

The current progressive Dragon's Head in Sagittarius (FOREIGNERS) will enhance your 7th house (PARTNERSHIPS / MARRIAGE) and stimulate your creativity leading you to more recognition even fame if you associate with foreigners or aim for foreign grounds. Do not let the current trying Dragon's Tail in Gemini affect your judgment, this would bring obstacles and frustration altering your mental clarity, actions and wisdom, patience is the key. All Gemini are strongly advised to stay clear from any form of chemicals, drugs, alcohol and anti depressant. Many Unlucky Gemini may not make it and will slowly suffer mental depletion, heavy depressions leading to suicidal tendencies. This Gemini Dragon's Tail may also induce stress with siblings and danger with transportation. (i.e., *UK deadly chain reaction. At least 7 dead, 51 hurt in traffic 'fireball' in Britain.*)

Remember, the Dragon's priority is to help you reevaluate your own wisdom, your beliefs and upgrade to a higher perception of your physical and mental self. Indeed 2011 is an important year forcing you to expand mentally and fulfill your teaching mission. The current Dragon's Head in Sagittarius will usher stimulating desires to find a partner and will bring about great associations or even contacts from foreigners dedicated to help you as you help them. With it the opportunity to introduce a new spiritual fresh way at looking at education in the world at large. Many opportunities will be offered to you as you upgrade your perception to the world, your teaching, learning potential and how you could also participate in helping the world. The Dragon's Head in your marriage house, wants you to adjust and adapt while being productive all along bringing your gifts to the world. While the unlucky Gemini may suffer total exhaustion, taking on all the challenges with patience will make you more efficient in reaching your wishes and avoid depressions. The month of June 2012 will be particularly trying for all of those born in June while the lucky Gemini will benefit from good karma. Right on your 7th house the blessings of the Dragon's Head will induce wonderful opportunities to reach the people you need most, travel teach and be even published. The Dragon is demanding of you to adjust to a new form of thinking, learning, teaching to stimulate your career and reach for a higher wisdom for better mental strength. Be ready to accept those changes with confidence because the future has a lot to offer you.

Until August 30th 2012, matters involving traveling, learning, teaching publishing and foreigners will become powerful driving forces for Gemini. Deserving souls born in June will benefit from foreign friends / teachers promoting many of their wishes. Many hard-working Gemini souls will get plenty of opportunities to expand following good business decisions. Some traveling opportunities will offer you a new understanding and a higher position in the world. The Dragon's Tail location in the sign of Gemini afflicting your entire being could bring havoc if you give in to an uncontrollable imagination, rush or try to too hard. Use your will and avoid negative thoughts when dealing with all the stress and fears the Tail of the Dragon may challenge you with. The disturbing Tail of the Dragon does not have to win over you but will induce serious mental challenges where patience becomes the key. Many Gemini souls may find themselves rebuilding their entire public life and this could stir fears of the future. Souls born in June will need to adjust to a new way of thinking and if you feel the Dragon's depressive, panicking impact you MUST contact me and let me work on your spirit. One hour on the telephone or on Skype will do the job and bring back hope, health, faith and magic into your life. (602) 265-7667

2012 Predictions:
On August 30th 2012, the Dragon changes axis from Sagittarius / Gemini into Scorpio / Taurus.

The new Dragon's Head (LUCK / GROWTH) will stimulate your 6th house (WORK AND HEALTH) where you will experience a rebirth of your own aptitude to survive the world. The Dragon will induce a new discerning healing power and with it an enormous surge of ideas leading to financial creativity. This impact will upgrade your physical health and sense of security working within the structure of powerful well established groups. The difficult dragon energy will affect your spiritual life inducing a full serious psychological changes and demand you to constantly regenerate with creative or investigative material. Souls born in June will be tested to the limits in the affairs involving the subconscious making them very vulnerable to depressions, addictions and even suicide. The year 2012 could mark the beginning of an era that could mean the physical and spiritual rebuilding of life or the loss of it all for many Gemini. Others will enjoy a well deserved, well respected position. The lucky Gemini soul will be able to manage both the physical and spiritual realms promoting them towards a successful year. Many deserving souls born in June will begin to see the results of many years of arduous mental work where more opportunities will flourish.

JUPITER LUCKY TOUCH – In 2012, the great beneficial planet Jupiter (LUCK / EXPANSION / PROTECTION / TRAVELING / FOREIGNERS / STUDYING) will be cruising though the sign of Taurus (MONEY) until June 12th 2012. Jupiter's luck will add benefit to the 12th house affairs related to their spiritual life. The challenging effect produced by the Tail of the Dragon (KARMA) in Taurus will be seriously altered with Jupiter's protection. With Jupiter's blessings in Taurus affecting all financial institutions (BANKS) it could also bring help from foreign countries. This impact could also be perceived as negative as foreign countries are buying / owing more and more prime real estate, offering them the option to invest and make a lot of money in the process. When Jupiter (LUCK) moves in Gemini on June 12th 2012, expect an upsurge of information pertaining to banks restructures coming your way. You could also benefit and invest on foreign ground in the process. Souls born in June should be ready for beneficial career associations offering great prospective business deals with souls born in February, October, December and those born in April. Working in good knowledge of your "Personal Lucky Dragon Window Dates" will also become a serious contribution to save you time especially if you do a lot of traveling. Being at the right place at the right time has a lot to do with your progress in terms of lucky breaks and opportunities. In 2012, many Gemini will be forced to reevaluate the way they handle their inner life and their health. Many Gemini will also be offered (*or will chose*) the option to relocate and reach a much bigger world in the process; The knowledge found in Astro-Carto-Graphy would be a major contribution to your success (*or your failure*) in one of these new locations. Keep this opportunity in mind and give it a try — it works! http://www.drturi.com/readings/ - Good luck to all souls born in June.

Cancer
Philosophical Astro – Poetry

The Moon Governs the Nurturing and Caring Constellation of Cancer

I am mother I nurture and provide
In my soul the physical and spiritual collide
I say, "ask and you shall receive."
But also "as you sow, so shall you reap"
I am Cancer, child of the Moon.

Characteristics For Those Born In July

The moon and the emotional sign of Cancer rule your life, as a moonchild your life is strongly affected by the Moon's fluctuations. Knowing and using her whereabouts in the belt of the Zodiac is a must for you to succeed in all areas of your existence. Family matters will always play an important part in your life. You are classified as the "caretaker" of the Zodiac. Much of your success depends on the awareness and ability to use both the moon and your powerful intuition. You are distinctively gifted with real estate and food (*cooking or eating*). Lunar children have a solid sense of organization and have inherited strong managerial gifts.

You are a perfectionist and are quite critical at times. Plants and green appeal to you, and you tend to worry too much about health to the point of becoming a vegetarian. You will perform very well in a position of power or management. Financial security is important to you and you will shine through your ability to amass riches and possessions, (*i.e., Real estate mogul, Ross Perot*). You have a gift with children and you have a natural zest to teach them. You must avoid depressing thoughts of the past and keep control over your powerful imagination. Steadiness, organization, warmth, love, and charm belong to you. Your powerful emotions can be channeled positively with music, singing, and the arts in general (*country music is a Cancer / July vibration*).

You are attracted to successful people (*older or younger mates*) and many Cancerian want to marry rich. Your natural tendency to smother family members and friends at all times makes you admired and deeply loved. You must learn to control your overwhelming sensitivity and participate with life outside of your home a little more. As a rule, all Moon children are great homemakers unless the soul selected a non domestic masculine Moon before reincarnating on this dense physical world. Like all other water signs you regenerate in research, science, and metaphysics. You tend to worry too much about your and others health and you should adopt a more positive spiritual attitude. Learn to let go of the wrong people and move on with life. It is a must for you to respect the Universal Law (*see Moon Power*), as your awareness and moon planning will become a major contribution toward avoiding dramatic experiences and reaching many of your dreams. The location of your natal Dragon's Head or Tail will seriously alter the strengths or weakness of the Moon in your chart. You can learn much more about yourself or anyone else by ordering my new book entitled *"Beyond The Secret," "I Know All About You," "The Power Of The Dragon"* or *"And God Created The Stars."*

2012 — Dragon Forecast For Those Born in July

Personal: On March 4th 2011, the karmic Dragon moved in the Axis Sagittarius *Head / Gemini *Tail and will stay there until August 30th 2012, stimulating your 6th and 12th houses regulating your subconscious creative / destructive forces and your work and service to the world. Then in

August 30th 2012, the Dragon changes its axis from Sagittarius / Gemini into Scorpio / Taurus and will stimulate all affairs regulating your wishes, friends, groups and your love, children, romance and speculations area.

On your 12th house (SUBCONSCIOUS) the difficult Dragon's Tail will force you to undergo serious psychological changes making you very vulnerable to any induced fears. Some may be apocalyptic and based upon religious material other from reading/listening to negative materials. This impact will force you "rebuild" your spirit appropriately with progressive spiritual material or suffer the consequences of uncontrolled imagination that could lead to serious depressions and even suicidal tendencies. The progressive Dragon's Head in Sagittarius will induce healing spiritual material generated from foreigners. Feeding your spirit with good food will rebuild your physical atomics structure leading you to exercise or undergo worthwhile dieting programs and present a new you and a new image to the world. Indeed a very important year for you as you must be "reborn" in both the physical and spiritual world. The Dragon's Head will induce a new spiritual fresh way at looking at the world and how you could participate helping others in a big way. The Dragon's Tail on your mind will play havoc with your imagination while the unlucky Cancer may suffer serious depressions. The months of June and December 2011 will be particularly trying for *All* of those born in July. Right on your mental health and work area, the Dragon will induce serious challenges forcing you to reevaluate what might hurt you physical and mentally. The Dragon is demanding of you to adopt a new spiritual way to perceive yourself in the world and may lead you to take on new studies. Be ready, and accept those changes with confidence and let go of the old fearful imaginative self.

Until August 30th 2012, matters involving the spirit, religions, health, and education will become powerful driving forces in 2011. The lucky souls born in July will enjoy a spiritual uplifting and a great dose of well controlled creative imagination. Many hard-working souls born in July will get the opportunity to improve their physical and spiritual images leading them to initiate good business decisions. Following a higher study many Cancer souls will be promoted to a new higher position to service the world. This will open exciting doors with new friends and traveling will lead to more worthwhile business partnerships. The Dragon's Tail location in the sign of Gemini in your subconscious house could bring depressions if you allow fears of the future to afflict your faith and your spirit. Use your will and avoid disheartening thoughts when dealing with all the stress and fears the Tail of the Dragon may challenge you with. The disturbing Tail of the Dragon is inducing serious challenges and will force you to reevaluate your accepted spiritual convictions and realize the importance of neutralizing and eliminating any and all form of poisoning fears. Many unlucky Cancer souls may find spiritual relief with legal / illegal drugs or alcohol, making the situation much worse. More than any other signs of the zodiac, in 2011, souls born in July will need a serious dose of constant spiritual regeneration to counteract the relentless Dragon, if you feel its depressive, panicking impact you MUST contact me and let me work on your spirit. One hour on the telephone or on Skype will do the job and bring back hope, health, faith and magic into your life. (602) 265-7667

2012 Predictions:
On August 30th 2012, the Dragon changes axis from Sagittarius / Gemini into Scorpio / Taurus.

The new Dragon's Head (LUCK / GROWTH) will affect your 5th house of speculation, love, romance and children where you will experience serious progression and many of your wishes coming true. This impact may affect your sense of security and bring hopes in the affairs of love, romance and children. The new dragon energy may stir a new sense of faith where the drive for new adventures and speculations will open great doors. Souls born in July will be tested to the limits in the affairs involving security, money, love and children. Indeed, 2012 could mark the end of important relationships that will directly affect their children and the support to them. The lucky Cancer soul will be able to manage both and be greatly rewarded joining others to reach emotional, financial and spiritual stability. Many souls born in July will be starting, joining or leaving groups or associations or be forced to adjust to karmic unseen metaphysical forces.

JUPITER LUCKY TOUCH – In 2012, the great beneficial planet Jupiter (LUCK / EXPANSION / PROTECTION / TRAVELING / FOREIGNERS / STUDYING) will be cruising though the sign of Taurus (MONEY) until June 12th 2012. Jupiter's luck will benefit you in the affairs involving groups organization seriously altering the nasty effect produced by the Tail of the Dragon (KARMA) in this house. With Jupiter's blessings on your 11th house of wishes, new friends will bring wisdom and the light needed to succeed. The new found wisdom will bring the option to invest and make a lot of money (*or wisdom*) in the process. With Jupiter's blessings in Taurus affecting all financial institutions (BANKS) it could bring help from foreign countries. This impact could also be negative as foreign countries are buying/owing more and more real estate, offering the option for them to invest and make a lot of money in the process. When Jupiter (LUCK) moves in Gemini on June 12th 2012, expect more traveling and a wave of new banks restructures and information coming your way. You will enjoy foreign people, foreign ground and foreign food packing up extra pounds (*or serious wisdom*) in the process. Souls born in July should be ready for new associations and great business deals with those born in November, January, March and May. Working in good knowledge of your "Personal Lucky Dragon Window Dates" will also become a serious contribution to saving you time and money especially if you do a lot of traveling. Being at the right place at the right time has a lot to do with your progress in terms of lucky breaks and opportunities. In 2012, many souls born in July will be forced to reevaluate their work, health and service to the world. Many Cancer will be offered (*or will chose*) the option to relocate in the process; The knowledge found in Astro-Carto-Graphy would be a major contribution to your success (*or your failure*) in one of these new locations. Keep this opportunity in mind and give it a try — it works! http://www.drturi.com/readings/ - Good luck to all souls born in July.

Leo

Philosophical Astro – Poetry

The Sun Governs the Flamboyant and Majestic Constellation of Leo

Powerful and Charming
All things living find me disarming
I step to the center of God's stage
In the books of history I have always a page
I am Leo, child of the Sun.

Characteristics For Those Born In August

The month of August is governed by the all-powerful Sun and by the magnanimous sign of Leo. Your solar sign reflects the dignified Sun's life force energy and classifies you as "The Life Giver." During the day the Sun outshines all the other planets giving you the option to reach fame, fortune and power and shine in the stage of life. Naturally gifted, you are attracted to professions involving the arts, public life, medicine, research, management, and any endeavors that could offer you a chance to shine. (*i.e., President Obama, Madonna, Michael Jackson, Schwarzenegger to name a few.*) Just as the Sun's rays penetrate the depths of the rainforest, you were born with the potential to bring and promote life to all that you touch. You have a lot to offer others and the world, providing you exercise control over your ego and your authoritative nature. The untamed "King of the Jungle" must positively direct and control the Sun's creative force without burning himself or others in the process. You are fixed and strongly motivated by the will to succeed. Strong and dominant, you nurture a formidable desire to organize and rule others. If you become too overbearing, others will then teach you the lesson of humility where you will be forced back down and start from scratch. Destructive outbursts of emotions and unfettered pride are enemies of success.

Your challenge is to recognize the powerful Sun's energy and diligently work towards a better understanding and respect of others. Acting eccentrically or with pride and without forethought is your weakness. However you will courageously handle all the difficulties of life. The advanced Leo possesses nobility of purpose and great spiritual values. Women born in August are stunning, intellectual, magnetic, and attract others with their enthusiastic solar power. Women born in August are also protective and dedicated mothers. The desire for fame could also make them overbearing and try to live through their children's accomplishments. The Sun rules life and you may nurture a subconscious fear of death and decay. But nature gives you a strong mind and a robust body. You love animals, especially horses. You tend to be weak and accident prone in the back, knees and joint areas. (*President Clinton was born in August and, busted his knee in Florida!*) A word of caution for those born in August: use precaution and moderation when running or jogging. Remember to respect the Universal Law (*see Moon Power*), as your awareness of Sun / Moon lucky combined with your personal lucky window dates will become a major contribution to success. The location of your natal Dragon's Head or Tail will seriously alter the strengths or weakness of the Sun in your chart. You can learn much more about yourself or anyone else by ordering my new book entitled *"Beyond The Secret"* or *"I Know All About You," "The Power Of The Dragon"* or *"And God Created The Stars."*

2012 — Dragon Forecast For Those Born in August

Personal: On March 4th 2011, the karmic Dragon moved in the Axis Sagittarius *Head/ Gemini *Tail and will stay there until August 30th 2012, stimulating your 5th and 11th houses regulating love, romance, children, speculations and your 11th house of groups, wishes and friends. Then in August 30th 2012, the Dragon changes its axis from Sagittarius / Gemini into Scorpio / Taurus and will stimulate all affairs regulating your public standing and your home life area.

In your 11th house (GROUPS / FRIENDS) the trying Dragon's Tail *negative will offer you great opportunities to reach a bigger and higher place in the world only if you adjust to the rules and demands of others. Doing so will enable you to operate and to be part of a group making you very much in demand to reach a bigger world. This impact will force you to "rebuild" a new, more diplomatic you with great progressive material to share with others; this could lead you to more recognition even fame. The Dragon's Tail in Gemini could bring mental stress, being involved in so many projects. This Dragon may also induce stress with loved ones or children wanting more of your attention. Remember your priorities and return love to those who need it most. Indeed a very important year for you where great opportunities will be offered by foreigners to expand to fulfill your goals and reach for all corners of the world. The current Dragon's Head in Sagittarius will induce great contacts from foreigners or from foreign grounds, with it the opportunity to introduce a new spiritual fresh way at looking at yourself and how you fit in the world. Many opportunities will be offered to you on how you could participate helping others if you listen more. The Dragon's Tail in your friends and wishes house demands you to adjust and adapt while being productive all along and to re-create a new image to the world. While the unlucky Leo may lose great friends or suffer relationships challenges, patience and understanding is the key. The month of June 2012 will be particularly trying for all of those born in August while the lucky one will benefit from good karma. Right "on your love, romance, children and creativity house the Dragon's Head will induce serious personal challenges and eliminate whom ever, what ever stands on your way to success. The Dragon is demanding of you to join others for strength without hurting those you care. Be ready to accept those changes with confidence be productive all along.

Until August 30th 2012, matters involving group association, traveling, publishing and foreigners, including love, romance and children will become powerful driving forces. Deserving souls born in August will enjoy new friends promoting many of their wishes and with it the option to find love. Many hard-working Leo souls will get plenty of opportunities to expand, learn and teach following good business decisions. Some traveling will also offer you with new and valuable promotions and a better higher position to reach the world. The Dragon will open exciting new doors to reach the world leading friends to engage in worthwhile business partnerships. The Dragon's Tail location in the sign of Gemini in your 11th house (FRIENDS / WISHES) could bring havoc if you rush or miss their aims. Use your will and avoid negative thoughts when dealing with all the stress and fears the Tail of the Dragon may challenge you with. The disturbing Tail of the Dragon does not have to win over you but will induce serious challenges with friends and groups including loved ones forcing you to adapt. Many unlucky Leo souls may find freedom at the expenses of those they care just to realize their error in the future. Souls born in August will need to think before acting and hurt those they care for. If you feel its depressive, panicking impact you MUST contact me and let me work on your spirit. One hour on the telephone or on Skype will do the job and bring back hope, health, faith and magic into your life. (602) 265-7667

2012 Predictions:

On August 30th 2012, the Dragon changes axis from Sagittarius / Gemini into Scorpio / Taurus.

The new Dragon's Head (LUCK / GROWTH) will stimulate your 4th house (HOME / FAMILY / REAL ESTATE) where you will experience a serious expansion for security. This impact will upgrade your sense of security for your family and may force you to move or expand the base of your operations. (*Obama born in August will be forced to exit the White house...The new dragon energy will also stir Mr. Ego and Mrs. Pride where reality must be accepted.*) Souls born in August will be tested to the limits in the affairs involving career, home and security but 2012 could mark the beginning of an era that will directly affect their position in the world. The lucky Leo soul will be able to manage both a very busy career and home responsibility leading to emotional, financial and spiritual stability. Many deserving souls born in August will see the results of many years of arduous work where more opportunities will flourish.

JUPITER LUCKY TOUCH – In 2012, the great beneficial planet Jupiter (LUCK / EXPANSION / PROTECTION / TRAVELING / FOREIGNERS / STUDYING) will be cruising though the sign of Taurus (MONEY) until June 12th 2012. Jupiter's luck will benefit you in the affairs involving your base of operation, real estate, family matters seriously altering the nasty effect produced by the Tail of the Dragon (KARMA) in this house. With Jupiter's blessings in Taurus affecting all financial institutions (BANKS) this could bring help from foreign countries. This impact could also be perceived as negative as foreign countries are buying / owing more and more prime real estate, offering the option for them to invest and make a lot of money in the process. When Jupiter (LUCK) moves in Gemini on June 12th 2012, expect more traveling and a wave of information pertaining to banks restructures coming your way. You will enjoy your own opportunities to invest on foreign ground in the process. Souls born in August should be ready for great career associations and great business deals with souls born in December, February, April and those born in June. Working in good knowledge of your "Personal Lucky Dragon Window Dates" will also become a serious contribution to saving you time and money especially if you do a lot of traveling. Being at the right place at the right time has a lot to do with your progress in terms of lucky breaks and opportunities. In 2012, many souls born in August will be forced to reevaluate their career and security. Many Leo souls will also be offered (*or will chose*) the option to relocate in the process; The knowledge found in Astro-Carto-Graphy would be a major contribution to your success (*or your failure*) in one of these new locations. Keep this opportunity in mind and give it a try — it works! http://www.drturi.com/readings/ - Good luck to all souls born in August.

Virgo

Philosophical Astro – Poetry

Mercury Governs the Precise and Critical Constellation of Virgo

Cleansing impurities large and small
Don't think yourself immune, for I see all
Attending to every chore and task
Perfection being all that I ask
I am Virgo, child of Mercury.

Characteristics For Those Born In September

The month of September is governed by the planet Mercury and by the critical sign of Virgo. You are an intellectual sometime quite critical, even picky with others and you tend to work much too hard. You were born a master of communications endowed with a deep analytical thinking process; the stars also offer you the option to become a great speaker and a proficient writer. When challenged you may misuse this power and become sarcastic to others. You will always combine logic and intuition in dealing with life in general. Astrologically, you have been classified as the "perfectionists" and your editing touch is satisfied only with perfection.

You will do well in the fields of technology, medicine, law, investigation, teaching, writing, designing, and office work and you are also a refined artist. Your downfall is sarcasm and an overly concerned attitude with trivial matters. Some young Mercurial souls are overwhelmed with health matters and turn themselves into health fanatics. Other Virgos simply refuse to work hard and succumb to chemical, drugs and alcohol abuses. Letting the rational mind scrutinize everything can hinder your spiritual gifts and neutralize your cosmic consciousness. You have inherited a powerful investigative mind that could lead you to science, chemicals, research, radio, television, newspaper reporting, computer programming and the law.

Spiritually advanced September souls are born mental leaders and masters in communication, (i.e., Dr. Drew, Robert Shapiro and Marcia Clark {O.J. Simpson trial attorneys}) are Virgos and indicative of your intellectual potential pertaining to medicine, investigations and the law. You may also be prone to headaches or head injury, eyes and sinus migraine problems. Be aware of your environment in public places. You are prone to poisoning and strongly advised to keep away from alcohol and narcotics, (i.e., Michael Jackson). Also, be aware, diets that are too restrictive may cause just as many problems as over-indulgence. Your body and metabolism are both well equipped to deal with all types of food, including red meat. If your natural desire for perfection prevails and you eliminate this "red" source of food, you must then substitute it with different red foods such as red wine, hot peppers or other thermogenic foods. If you happen to suffer headaches or migraines, you may find relief by walking barefoot on the grass (*or close to a body of water*) to regenerate from the earth's magnetic field. As a rule all souls born in September are sadly enough, accident prone to head trauma and violent death. Therefore, if you were born in September, do not take chances, especially during or after the Full Moon or during your unlucky Dragon window dates. Keep in mind to respect the Universal Law (*see Moon Power*), as your awareness of the Mercury / Moon planning will become a major contribution for success and happiness. The location of your natal Dragon's Head or Tail will seriously alter the strengths or weakness of Mercury in your chart. You can learn much more about yourself or anyone else by ordering my new book entitled *"Beyond The Secret," "I Know All About You," "The Power Of The Dragon"* or *"And God Created The Stars."*

2012 — Dragon Forecast For Those Born in September

<u>Personal:</u> In March 4th 2011, the karmic Dragon moved in the Axis Sagittarius *Head / Gemini *Tail and will stay there until August 30th 2012, stimulating your 4th and 10th houses regulating general security and career matters. Then in August 30th 2012, the Dragon changes its axis from Sagittarius / Gemini into Scorpio / Taurus and will stimulate all affairs regulating your mind, higher learning's and publishing.

In your 4th house (HOME / REAL ESTATE) the Dragon's Head *positive will offer you great opportunities to relocate to a dryer or suburban area much closer to offering you the option to enjoy calmness and the wild. There you will be able to channel more from the Indian spirits fulfilling your endless drive to educate others. This impact will force you to "rebuild" a new you and still reach the world electronically. Your hard work and drive could lead you to more recognition even fame if you associate with foreigners or aim for foreign grounds. The Dragon's Tail in Gemini keeps bringing obstacles related to universally publishing your wisdom inducing more frustrations, patience is the key. This Dragon may also induce stress with siblings and transportation. Remember, the Dragon's priority will be forcing you to reevaluate your wisdom, beliefs and upgrade to a higher wisdom. Indeed a very important year for you being forced to expand mentally to fulfill your mission by reaching a much bigger world. The current Dragon's Head in Sagittarius will bring about association or contacts from foreigners dedicated to help you growth spiritually, with it the opportunity to introduce a new spiritual fresh way at looking at education in the world. Many opportunities will be offered to you as you progress on how you could also participate in helping others. The Dragon's Tail in your career area wants you to adjust and adapt while being productive all along bringing your gifts to the world. While the unlucky Virgo may suffer a total career collapse, taking on the challenges with patience will make you more efficient in reaching the world. The month of June 2012 will be particularly trying for all of those born in September while the lucky Virgo will benefit from good karma. Right on your public standing house, the Dragon's Tail will induce serious personal challenges. The Dragon is demanding of you to adjust to a new home and career life join and join others for strength. Be ready to accept those changes with confidence the future has a lot to offer you.

Until August 30th 2012, matters involving career and group association, traveling, publishing and foreigners will become powerful driving forces for Virgo. Deserving souls born in September will enjoy a new base of operation and new friends promoting many of their wishes. Many hardworking Virgo souls will get plenty of opportunities to expand, learn and teach following good business decisions. Some traveling will offer you with new and valuable opportunities and with time, a better higher position in the world. The Dragon will open exciting new doors leading many of your friends to become involved in business partnerships. The Dragon's Tail location in the sign of Gemini in your 10th house (CAREER) could bring havoc if you rush or try to hard. Use your will and avoid negative thoughts when dealing with all the stress and fears the Tail of the Dragon may challenge you with. The disturbing Tail of the Dragon does not have to win over you but will induce serious restructuring challenges where patience is the key. Many Virgo souls may find themselves rebuilding their entire life and could stir fears of the future. Souls born in September will need to adjust to a new life and if you feel the Dragon's depressive, panicking impact you MUST contact me and let me work on your spirit. One hour on the telephone or on Skype will do the job and bring back hope, health, faith and magic into your life. (602) 265-7667

2012 Predictions:
On August 30th 2012, the Dragon changes axis from Sagittarius / Gemini into Scorpio / Taurus.

The new Dragon's Head (LUCK / GROWTH) will stimulate your 3rd house (COMMUNICATIONS) where you will experience a rebirth of your consciousness, this will induce tremendous intellect discerning power and an enormous surge of creativity. This impact will upgrade your spiritual senses and induce channeling and a urge for writing. The new dragon energy will also the Plutonic energy demanding you to constantly regenerate with creative or investigative material. Souls born in September will be tested to the limits in the affairs involving the mind, learning, teaching, publishing inducing traveling. The year 2012 could mark the beginning of an era that could lead them to a teaching well respected position. The lucky Virgo soul will be able to manage both the physical and spiritual realms promoting them towards a very busy career. Many deserving souls born in September will begin to see the results of many years of arduous mental work where more opportunities will flourish.

JUPITER LUCKY TOUCH – In 2012, the great beneficial planet Jupiter (LUCK / EXPANSION / PROTECTION / TRAVELING / FOREIGNERS / STUDYING) will be cruising though the sign of Taurus (MONEY) until June 12th 2012. Jupiter's luck will add benefit to the 9th house affairs related to intellectual properties, teachings and publishing. The challenging effect produced by the Tail of the Dragon (KARMA) in the 9th house will be seriously altered with Jupiter's protection. With Jupiter's blessings in Taurus affecting all financial institutions (BANKS) it could bring help from foreign countries. This impact could also be negative as foreign countries are buying / owing more and more prime real estate, offering the option to invest and make a lot of money in the process. When Jupiter (LUCK) moves in Gemini on June 12th 2012, expect an upsurge in your career inducing more traveling and a wave of information pertaining to banks restructures coming your way. You could also benefit and invest on foreign ground in the process. Souls born in September should be ready for mental exploration and beneficial career associations offering great prospective business deals with souls born in March, May, January and those born in July. Working in good knowledge of your "Personal Lucky Dragon Window Dates" will also become a serious contribution to saving you time and money especially if you do a lot of traveling. Being at the right place at the right time has a lot to do with your progress in terms of lucky breaks and opportunities. Many Virgo's will be forced to reevaluate their spiritual values, career matters and security. In 2012, many Virgo's will also be offered (*or will chose*) the option to relocate in the process; The knowledge found in Astro-Carto-Graphy would be a major contribution to your success (*or your failure*) in one of these new locations. Keep this opportunity in mind and give it a try — it works! http://www.drturi.com/readings/ - Good luck to all souls born in September.

Libra

Philosophical Astro – Poetry

Venus Governs the Diplomatic and Peaceful Constellation of Libra

Lover of grace and harmony
Seeking the balance of matrimony
Though there are those that hold to opinions tight
I will see it in all the different lights
I am Libra, child of Venus.

Characteristics For Those Born In October

The month of October is ruled by the planet Venus and by the charming sign of Libra. You are strongly motivated by a desire for justice and you must create harmony in all areas of life. You are classified as the "Peacemaker" in Astropsychology. You will succeed in your career because of your gentle personality, your sense of diplomacy and your natural "savoir faire." You rarely learn by mistake, but you must avoid prolonged indecision. Those born in October must establish Libra's soul's purpose of achieving balance, emotional, financial, and spiritual stability during the course of their lifetimes. You must stand for yourself and learn decision making by following not only your rational mind but also your accurate intuitions. You possess a strong psychological aptitude and do well in the real estate and the food industries, the stock market, interior design, marriage counseling and the arts in general. You must focus on what you need first by using inner stamina and both your practical and intuitive minds. These gentle personalities will be attracted to competitive people and one can expect many challenges from them. Rough behavior or the abrupt and assertive manner of a business partner easily offends Libras, this nefarious attitude transform the gentle Libra soul into a very aggressive, violent, abusive person forced to learn the hard way about the law. The same desire for diplomacy is expected from a friend or a lover or simply expect the worse from them. Libra should also avoid taking remarks too personally and use self control when challenged. Libra's love a good home and you enjoy the company of business-oriented partners.

Your downfall comes from traditional scientific or religious teachings and your refusal to challenge early religious tutoring or addictions. Casting aside self-discovery and real spirituality will slow down or eliminate your chances to develop your cosmic consciousness and establish emotional, financial and spiritual stability. Libra are very smart, well read and eternal students, all are born great students. This codification of thoughts produces a crowd of educated mental snobs, librarians, ministers, priests and religious leaders. As indicated by Libra's scale, both human and divine Universal laws must be acknowledged and respected. Using both traditional and untraditional means of education will bring about a better awareness of these laws. The limitation of conventional education (*psychology or religion*) is overridden by a more progressive spiritual attitude (*New Age and Astropsychology*) and will bring about all the answers you seek. You are a philosopher and a great teacher; likewise, you will travel far in search of the truth. The truth you are aiming for is right above your heads in the stars. A word of caution to you: stay clear of all chemicals, such as pot, drugs, or alcohol, as these destructive habits could blur your spirit or worse lead you to a hospital or jail. Remember to respect the Universal Law (*see Moon Power*), your awareness of the Universal Laws will become a major contribution for your happiness and success in life. The location of your natal Dragon's Head or Tail will seriously alter the strengths or weaknesses of Venus in your chart. You can learn much more about yourself or anyone else by ordering my new book entitled *"Beyond The Secret," "I Know All About You," "The Power Of The Dragon"* or *"And God Created The Stars."*

2012 — Dragon Forecast For Those Born in October

<u>Personal:</u> On March 4th 2011, the karmic Dragon moved in the Axis Sagittarius *Head / Gemini *Tail and will stay there until August 30th 2012, stimulating your 9th and 3rd houses regulating your mental power your higher learning potential and publishing. Then in August 30th 2012, the Dragon changes its axis from Sagittarius / Gemini into Scorpio / Taurus and will stimulate all affairs regulating your 2nd house (MONEY) and your 8th house (CORPORATE ENDEAVORS).

The challenging Dragon's Tail will impose a restructure of your higher learning potential forcing you to reevaluate your philosophical or religious values where foreigners or traveling foreign grounds could stimulate your drive for new mental exploration. You will need to be cautious gathering spiritual ungrounded material or the drive to educate others with what could be an illusive deceptive newfound religious wisdom. The pressing Dragon's Tail afflicting your learning potential could brings outside interferences from your past and induce uncontrolled imagination fuelled by fears of personal or career failures. Be cautious of all material you hear, read and use your highly intelligent nature to challenge the books instead and decipher its deceptive material. The current progressive Dragon's Head in Sagittarius will enhance your 3rd house (CRITICAL THINKING) and stimulate a more objective, spiritual attitude leading you to learn more about religions and humanity's hijacked spirit. You will own the golden keys of what it means to be human if you associate with foreigners or aim for foreign educational, logical material explaining the working of the Divine and God' signs. The Dragon's Tail in Gemini in your higher learning study area could bring obstacles and frustration altering your mental clarity, actions and wisdom, patience is the key.

Remember, the Dragon's priority is to help you reevaluate your wisdom, beliefs and upgrade to a higher perception of your spiritual self. Indeed 2011 is an important year forcing you to expand mentally and fulfill your own learning / teaching mission. The current Dragon's Head in Sagittarius will bring about great association or even contacts from foreigners dedicated to help you to perceive the world more objectively and without fears. With it the opportunity for you to learn and introduce a new spiritual fresh way at looking at education in the world at large. Many opportunities will be offered to you as you upgrade your spiritual perception and your teaching, learning potential and how you could also participate in helping the world. The Dragon's Tail *negative in your traveling house and learning / teaching area wants you to adjust and adapt while being productive all along bringing your own gifts to the world. While the unlucky Libra may suffer depressions and confusions, taking on all the challenges with patience will make you more efficient in reaching the true spirit and your own salvation. The month of June 2012 will be particularly trying for all of those born in October while the lucky Libra will benefit from good karma. Right on your mental houses the blessings of the Dragon's Head and Tail will induce wonderful opportunities to learn, travel, teach and be even published. The Dragon is demanding of you to adjust to a new form of philosophy that will stimulate your career and reach for a higher wisdom and more mental strength. Be ready to accept those changes with confidence because the future has a lot to offer you.

Until August 30th 2012, matters involving traveling, learning, teaching publishing and foreigners will become powerful driving forces for Libra. Deserving souls born in October will benefit from foreign friends / teachers promoting many of their wishes. Many hard-working Libra souls will get plenty of opportunities to expand following good business decisions. Some traveling opportunities will offer you a new understanding and a higher position in the world but be aware of the Dragon's Tail afflicting travels, your health and your security in foreign lands. The Dragon's Tail location in the sign of Gemini afflicting this house could bring havoc if you give in to an uncontrollable fearful imagination. Use your will and avoid negative thoughts when dealing with all the stress the Tail of the Dragon may challenge you with. The disturbing Tail of the Dragon does not have to win over you but will induce serious spiritual challenges where patience becomes the key. Many Libra souls may find themselves rebuilding their entire spiritual life and this could stir insecurity about the future. Souls born in October will need to adjust to a new way of perceiving God and if you feel the Dragon's depressive, panicking impact you MUST contact me and let me work on your spirit. One hour on the telephone or on Skype will do the job and bring back hope, health, faith and magic into your life. (602) 265-7667

2012 Predictions:
On August 30th 2012, the Dragon changes axis from Sagittarius / Gemini into Scorpio / Taurus.

The new Dragon's Tail in Taurus (BANKS) will stimulate your 8th house (CORPORATE MONEY / CONTRACTS / LEGACY) where you will experience a rebirth or a full difficult restructure of your finances. The Dragon will induce financial responsibility for others following a break up and with it unwanted stress. This impact will affect your self esteem and induce a new way of financial planning. The dragging dragon in your personal finances will also demand of you to constantly regenerate with creative or investigative spiritual material. Souls born in October will be tested to the limits in the affairs involving money. Abusive Libra souls will also feel the impact of an overdue karma with Uncle Sam or unfit partners where Libra is strongly advised to stay clean from criminal activity. The year 2012 could mark the beginning of an era that could mean financial security or the loss of it all for many Libra. The new Dragon's Head (LUCK / GROWTH) will stimulate your 2nd house (CORPORATE MONEY / CONTRACTS / LEGACY) where you will experience a rebirth of your own aptitude to generate or make an income for yourself. The Dragon will induce a new discerning financial power and with it an enormous surge of ideas leading to investments and financial creativity. This impact will upgrade your sense of security working within the structure of powerful well established groups. Other Libras will enjoy a well deserved, well respected position. The lucky Libra soul will be able to manage both the physical and spiritual realms promoting them towards a successful year. Many deserving souls born in October will begin to see the results of many years of arduous mental work where more opportunities will flourish.

JUPITER LUCKY TOUCH – In 2012, the great beneficial planet Jupiter (LUCK / EXPANSION / PROTECTION / TRAVELING / FOREIGNERS / STUDYING) will be cruising though the sign of Taurus (MONEY) until June 12th

2012. Jupiter's luck will add benefit to the 2nd house affairs related to all money making schemes. The challenging effect produced by the Tail of the Dragon (KARMA) in Taurus will be seriously altered with Jupiter's protection. With Jupiter's blessings in Taurus affecting all financial institutions (BANKS) could also bring help from foreign countries. This impact could also be perceived as negative as foreign countries are buying / owing more and more prime real estate, offering them the option to invest and make a lot of money in the process. When Jupiter (LUCK) moves in Gemini on June 12th 2012, expect an upsurge of information pertaining to banks restructures coming your way. You could also benefit and invest on foreign ground in the process. Souls born in October should be ready for beneficial career associations offering great prospective business deals with souls born in April, June, October and those born in August. Working in good knowledge of your "Personal Lucky Dragon Window Dates" will also become a serious contribution to save you time especially if you do a lot of traveling. Being at the right place at the right time has a lot to do with your progress in terms of lucky breaks and opportunities. In 2012, many Libra will be forced to reevaluate the way they handle money. Many Libra will also be offered (*or will chose*) the option to relocate and reach a much bigger world in the process; The knowledge found in Astro-Carto-Graphy would be a major contribution to your success (*or your failure*) in one of these new locations. Keep this opportunity in mind and give it a try — it works! http://www.drturi.com/readings/ - Good luck to all souls born in October.

Scorpio
Philosophical Astro – Poetry

Pluto Governs the Mighty Constellation of Scorpio "The Eagle or Lizard."

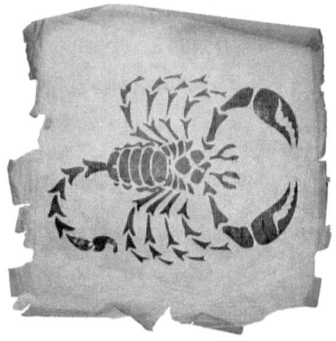

Holder of all the secrets deep
Never speaking for they are mine to keep
For those who plunder without care
Tread carefully for I see you there
I am Scorpio, child of Pluto.

Characteristics For Those Born In November

The planet Pluto and the intense sign of Scorpio govern the month of November. You inherited a powerful will and you are attracted to the unknown; the medical professions, the police force, metaphysics, politics, finances and general investigations. You are classified as the "Eagle" (*positive*) or the "Lizard" (*negative*) in Divine Astrology. You are quite private, secretive even mystic and like all other water signs you excel in the study and use of metaphysics. Unless you are aware of your innate witchcraft powerful residue, you are well advised not to sting yourself with your own dart. You carry in your soul the element of life and death, reincarnation and pure sensuality. On a negative note, your magnetic thoughts can reach anyone anywhere for good or for worse, bringing its accompanying karma into our own life. The young Scorpio soul will experience drama, despair and imprisonment during the course of his life. However, the destructive energies of Pluto can be channeled positively to accomplish tremendous results and incomparable career achievement. Your sign rules the Mafia, the police force, the FBI, CIA, supreme finance, and with it absolute power of creation or destruction, including sex. The message is quite clear when representing anyone born in November, deal with one and your life will change drastically. No one should take chances under Pluto's command. Realize that the Eagle in you is your challenge and your own birthright for creation or destruction.

These souls have no known fears in the face of death. Many advanced Pluto children will "fly" like an eagle above the destructive Lizard emotions and legendary jealousy. You can use your inborn mystical gifts to succeed where others would fail. Strong, private and dominant, you were born with a practical mind and an acute intuition. Your lesson is to control and direct constructively your deep emotions and use Pluto's ultimate power for the well being of society. You regenerate with investigations and spiritual matters and must uncover your unique mission in life. You are interested and aspire only for the undiluted truth and supreme power. The women of this sign are seen in Divine Astrology as "la femme fatale." They are sensual, classic, intellectual, reserved, and supremely magnetic. They also tend to use inner sexual power and physical beauty to reach their high goals. However, even as a powerful Scorpio, you are very weak with affairs of the heart and tend to be in love with love. A word of caution for you: Do not use your poisonous stinger against yourself or society. Remember to respect the Universal Law (*see Moon Power*), as your awareness and Moon planning will become a major contribution to your happiness and success. The location of your natal Dragon's Head or Tail will seriously alter the strength or weakness of Pluto in your chart. (*See Nostradamus Dragon Forecast for more information*). You can learn much more about yourself or anyone else by ordering my books entitled *"Beyond The Secret," "I Know All About You," "The Power of the Dragon"* or *"And God Created The Stars."*

2012 — Dragon Forecast For Those Born In November

<u>Personal:</u> On March 4th 2011, the karmic Dragon moved in the Axis Sagittarius *Head / Gemini *Tail and will stay there until August 30th 2012, stimulating your 8th and 2nd houses regulating your personal and corporate income. Then in August 30th 2012, the Dragon changes its axis from Sagittarius / Gemini into Scorpio / Taurus and will stimulate all affairs regulating your 1st house (YOURSELF) and your 7th house (MARRIAGE / CONTRACTS.)

The challenging Dragon's Tail will impose a restructure of your corporate finances and blur your self esteem forcing to adapt to the current depressive economy. Investigating new ways out, reading and writing will stimulate your endless creativity and the drive to reach for better solutions. You will need to stop worrying about losing more power and recharge your own mind positively. The mental challenges are simply exhausting as you want to help those you care deeply for. The difficult Dragon afflicting the house of death and metaphysics could bring about outside interferences from your past and induce uncontrolled imagination fuelled by fears of personal or career failures. The current progressive Dragon's Head in Sagittarius will enhance your 2nd house (MONEY) and stimulate creativity leading to new financial opportunities if you associate with foreigners or aim for foreign grounds. The Dragon's Tail in Gemini, in your corporate house, will bring obstacles and frustration altering your mental clarity, actions and wisdom; patience is the key. This Gemini Dragon's Tail may also induce stress with siblings, danger with transportation. (*i.e., UK deadly chain reaction. At least 7 dead, 51 hurt in traffic 'fireball' in Britain and even induce a legacy.*)

Remember, the Dragon's priority is to help you reevaluate your financial aptitudes and upgrade you to a higher perception of your creative self. Indeed, 2011 is an important year forcing you to expand mentally and fulfill your dream for endless financial security. The current Dragon's Head in Sagittarius will bring about great association or even contacts from foreigners dedicated to help you as you help them. With it the opportunity to introduce a new spiritual fresh way at looking at finances and education in the world at large. Many opportunities will be offered to you as you upgrade your perception of the divine and your learning potential and how you could also participate in helping the world to reach stability. The Dragon's Head in your personal money making house wants you to adjust and adapt while being productive all along bringing your own gifts to the world. While the unlucky Scorpio may suffer a total financial failure, taking on all the challenges with patience will make you more efficient in reaching your wishes and avoid depressions. The month of June 2012 will be particularly trying for all of those born in November while the lucky Scorpio will benefit from good karma. Right on your financial houses, the blessings of the Dragon's Head Tail will induce wonderful opportunities to travel and teach and be even published. The Dragon is demanding of you to adjust to a new form of thinking about possession and wealth while learning, teaching will stimulate your career. Be ready to accept those changes with confidence because the future has a lot to offer you.

Until August 30th 2012, matters involving finances, traveling, learning, teaching, publishing and foreigners will become powerful driving forces for Scorpio. Deserving souls born in November will benefit from foreign friends / teachers promoting many of their wishes. Many hardworking Scorpio's will get plenty of opportunities to expand, following good business decisions. Some traveling opportunities will offer you a new understanding and a higher position in the world. The Dragon's Tail location in the sign of Gemini afflicting your dying / rebirthing house could bring havoc if you give in to an uncontrollable imagination, rush or try to too hard. Use your will and avoid negative thoughts when dealing with all the stress and fears the Tail of the Dragon may challenge you with. The disturbing Tail of the Dragon does not have to win over you but will induce serious mental challenges where patience becomes the key. Many Scorpio souls may find themselves rebuilding their entire financial life and this could stir fears of the future. Souls born in November will need to adjust to a way of thinking and if you feel the Dragon's depressive, panicking impact you MUST contact me and let me work on your spirit. One hour on the telephone or on Skype will do the job and bring back hope, health, faith and magic into your life. (602) 265-7667

2012 Predictions:
On August 30th 2012, the Dragon changes axis from Sagittarius / Gemini into Scorpio / Taurus.

The new Dragon's Head (LUCK / GROWTH) will stimulate your 7th house (MARRIAGE / CONTRACTS / PARTNERSHIPS / LEGACY) where you will experience a rebirth of your own aptitude to generate or make an income with others. The Dragon will induce a new discerning financial power and with it an enormous surge of ideas leading to investments and financial creativity. This impact will upgrade your sense of security working within the structure of powerful well established groups or with worthwhile partners. The new dragon energy in your marriage area will open many doors especially with foreigners where love can bless your life. Souls born in November will be tested to the limits in the affairs involving marriages and karmic relationships that must come to an end. Abusive November souls will also feel the impact of an overdue karma with Uncle Sam while others are strongly cautioned to stay clean from criminal elements. The year 2012 could mark the beginning of an era that could mean enjoying or ending a long lasting relationships and get help financially. Others will enjoy a well deserved, well respected partner and a higher position in the world of finances. The lucky Scorpio soul will be able to manage both their business and emotional partnerships offering them a very successful year. Many deserving souls born in November will begin to see the results of many years of arduous mental work where more opportunities will flourish.

JUPITER LUCKY TOUCH – In 2012, the great beneficial planet Jupiter (LUCK / EXPANSION / PROTECTION / TRAVELING / FOREIGNERS / STUDYING) will be cruising though the sign of Taurus (MONEY) until June 12th 2012. Jupiter's luck will add benefit to the 7th house affairs related to partnerships and deals from foreign grounds. The challenging effect produced by the Tail of the Dragon (KARMA) in Taurus will be seriously altered with Jupiter's protection. With Jupiter's blessings in Taurus affecting all financial institutions (BANKS) to receive help from foreign countries. This impact could also be perceived as negative as foreign countries are buying / owing more and more prime real estate, offering them the option to invest and make a lot of money in the process. When Jupiter (LUCK) moves in Gemini on June 12th 2012, expect an upsurge of information pertaining to banks restructures coming your way. You could also benefit and invest on foreign ground in the process. Souls born in November should be ready for beneficial career associations offering great prospective business deals with souls born in May, March, July and those born in September. Working in good knowledge of your "Personal Lucky Dragon Window Dates" will also become a serious contribution to save you time especially if you do a lot of traveling. Being at the right place at the right time has a lot to do with your progress in terms of lucky breaks and opportunities. In 2012, many Scorpio's will be forced to reevaluate the way they handle partnerships and finances making it the main focus. November souls will also be offered (*or will chose*) the option to relocate and reach a much bigger world in the process; The knowledge found in Astro-Carto-Graphy would be a major contribution to your success (*or your failure*) in one of these new locations. Keep this opportunity in mind and give it a try — it works! http://www.drturi.com/readings/ - Good luck to all of those born in November.

Sagittarius
Philosophical Astro – Poetry

Jupiter Governs the Philosophical and Educated Constellation of Sagittarius

I have traveled the worldwide
With naught but the law on my side
Yearning for the higher knowledge
All of God's creation as my college
I am Sagittarius, child of Jupiter.

Characteristics For Those Born In December

The planet Jupiter and the sign of Sagittarius govern the month of December. You are a philosopher, a natural teacher and classified in Divine Astrology as a "Truth Seeker." You do well in learning or teaching computers, aeronautics, law, religion, communications, radio, and language. You are also attracted to holistic healing, animals, Indians and the world of sports. Your desire to travel to foreign lands is quite strong and will lead you to uncover many philosophies. Doing so will take you far away, giving you the option to return with incredible knowledge to teach to the rest of us. You were born with the gift of teaching and you will always promote a form of purity and organization in life. You can also do quite well in office work and you can be extremely organized in financial endeavors. You inherited a quick mind from the stars and you can keep up with anyone willing to discuss knowledge and philosophy. You need to realize the importance of education and you must focus on your chosen goals. Jupiter, "The Lord of Luck," will throw you many blessings in your life. With discipline and determination you have the potential to become a great speaker, produce interesting books and novels.

The young Sagittarius soul is too concerned with finances and must learn to give, so that he may receive help from the accumulated good karma. You must adapt to the saying, "to be a millionaire, you must act and think like one." Your sign rules the wilderness, the desert, and many disappeared Indian civilizations. This also represents some of your past lives with the Incas, the Sumerians and Atlantis where you had a position of spiritual power and full Cosmic Consciousness while dealing with extra terrestrials. A word of caution: Souls born with an overbearing Jupiterian energy must guard against the codification of thoughts (*books*) and biblical archaic materials; your lesson is to realize that God cannot be confined to any man-made buildings, deities or archaic doctrines. The advanced ones (*truth seekers*) will lead the rest of us towards the reality of God's manifestation through the stars, master and teach the Cosmic Code rules. Remember to respect the Universal Law (*see Moon Power*), as your awareness of Moon planning will become a major contribution to your happiness and success. The location of your natal Dragon's Head or Tail will seriously alter the strength or weakness of Jupiter in your chart. You can learn much more about yourself or anyone else by ordering my books entitled *"Beyond The Secret," "I Know All About You," "The Power Of The Dragon"* or *"And God Created The Stars."*

2012 — Dragon Forecast For Those Born In December

Personal: On March 4th 2011, the karmic Dragon moved in the Axis Sagittarius *Head / Gemini *Tail and will stay there until August 30th 2012, stimulating your 1st and 7th houses regulating your self discovery and your partnership / contract / marriage house. Then in August 30th 2012, the Dragon changes its axis from Sagittarius / Gemini into

Scorpio / Taurus and will stimulate all affairs regulating your 12th house (SUBCONSCIOUS) and your 6th house (SERVICE TO THE WORLD).

The challenging Dragon's Tail in Gemini imposes a restructure of your partnerships house where you may feel unwilling to deal with others or the world at large, this will seriously affects how other people sees you. Business or emotional much of your relationships will undergo a serious cleansing. Your need to educate others is strong but you will need to slow down, stop and recharge your own spirit at time. The mental challenges could become exhausting trying to bring the light to the world. The difficult Dragon afflicting your perception of partners could be wrong or brings about outside interferences from your past and induce uncontrolled imagination fuelled by fears of personal or career failures.

The current progressive Dragon's Head in Sagittarius (FOREIGNERS) enhance your 1st house (YOURSELF) and stimulate your creativity leading you to more recognition even fame if you associate with foreigners or aim for foreign grounds. Do not let the current trying Dragon's Tail in Gemini affect your judgment, this would bring obstacles and frustration altering your mental clarity, actions and wisdom, patience is the key. All Sagittarius are strongly advised to remember losing a friend means also losing wishes, most importantly stay clear from any form of chemicals, drugs, alcohol and anti depressant. Many unlucky Sagittarius must realize who they are with and if the partners do not fit the bill the Dragon will act as a vacuum cleaner. This Gemini Dragon's Tail may also induce stress with siblings and danger with transportation. (*i.e., UK deadly chain reaction. At least 7 dead, 51 hurt in traffic 'fireball' in Britain.*)

Remember, the Dragon's priority is to help you reevaluate your own wisdom, your beliefs and upgrade to a higher perception of your physical and mental self. Indeed 2011 is an important year forcing you to expand mentally and obtain new acquaintances and evaluate partnerships in order to fulfill your teaching mission. The current Dragon's Head in Sagittarius will bring stimulate a desire to look and feel better about yourself with the option to find a partner. New progressive associations will bring about worthwhile contacts from foreigners dedicated to help you as you help them. With it the opportunity to introduce a new spiritual fresh way at looking at yourself, your own education and your position in the world at large. Many opportunities will be offered to you as you upgrade your own perception and how you could participate helping the world. The Dragon's Head in yourself wants you to adjust and adapt while being productive all along bringing your gifts to the world. While the unlucky Sagittarius may suffer total mental exhaustion, taking on all the challenges with patience will make you more efficient in reaching your wishes and avoid depressions. The month of June 2012 will be particularly trying for all of those born in December while the lucky Sagittarius will benefit from good karma. Right on yourself the blessings of the Dragon's Head will induce wonderful opportunities to reach for a new you and the people you unknowingly need most. The Dragon is demanding of you to adjust to a new form of behaving, thinking, learning, teaching to stimulate your career and reach for a higher wisdom leading you to mental strength. Be ready to accept those changes with confidence because the future has a lot to offer you.

Until August 30th 2012, matters involving traveling, learning, teaching publishing and foreigners will become powerful driving forces for Sagittarius. Deserving souls born in December will benefit from foreign friends / teachers promoting many of their wishes. Many hard-working Sagittarius souls will get plenty of opportunities to expand following good business decisions. Some traveling opportunities will offer you a new understanding and a higher position in the world. The Dragon's Tail location in the sign of Gemini afflicting your entire being could bring havoc if you give in to an uncontrollable imagination, rush or try to too hard. Use your will and avoid negative thoughts when dealing with all the stress and fears the Tail of the Dragon may challenge you with. The disturbing Tail of the Dragon does not have to win over you but will induce serious mental challenges where patience becomes the key. Many Sagittarius souls may find themselves rebuilding their entire public life and this could stir fears of the future. Souls born in December will need to adjust to a new way of thinking and if you feel the Dragon's depressive, panicking impact you MUST contact me and let me work on your spirit. One hour on the telephone or on Skype will do the job and bring back hope, health, faith and magic into your life. (602) 265-7667

2012 Predictions:
On August 30th 2012, the Dragon changes axis from Sagittarius / Gemini into Scorpio / Taurus.

The new Dragon's Head (LUCK / GROWTH) will stimulate your 12th house (SUBCONSCIOUS FORCES) where you will experience a rebirth of your own psyche leading you to a better aptitude to survive the world. The Dragon will induce a new intuitiveness, prophetic dreams discerning intuition and inner healing power. This impact will also force you to deal or upgrade your physical health and sense of security working within the structure of powerful well established groups. The dragon energy will affect your spiritual life inducing a full and serious psychological changes and demand you to constantly regenerate with creative or investigative material. Souls born in December will be tested to the limits in the affairs involving the subconscious making them very vulnerable to depressions addictions and even suicide. The year 2012 could mark the beginning of an era that could mean the physical and spiritual rebuilding of the spiritual and physical life or the loss of it all for many Sagittarius. Others will enjoy a well deserved, well respected position. The lucky Sagittarius soul will be able to manage both the physical and spiritual realms promoting them towards a successful year. Many deserving souls born in December will begin to see the results of many years of arduous mental work where more opportunities will flourish.

JUPITER LUCKY TOUCH – In 2012, the great beneficial planet Jupiter (LUCK / EXPANSION / PROTECTION / TRAVELING / FOREIGNERS / STUDYING) will be cruising though the sign of Taurus (MONEY) until June 12th 2012. Jupiter's luck will add benefit to the 6th house affairs related to health and work. The challenging effect produced by the Tail of the Dragon (KARMA) in Taurus will be seriously altered with Jupiter's protection. With Jupiter's blessings in Taurus affecting all financial institutions (BANKS) it could also bring help from foreign countries. This impact could also be perceived as negative as foreign countries are buying / owing more and more prime real estate, offering them the option to invest and make a lot of money in the process. When Jupiter (LUCK) moves in Gemini on June 12th 2012, expect an upsurge of information pertaining to banks restructures coming your way. You could also benefit and invest on foreign ground in the process.

Souls born in December should be ready for beneficial career associations offering great prospective business deals with souls born in June, August, April and those born in September. Working in good knowledge of your "Personal Lucky Dragon Window Dates" will also become a serious contribution to save you time especially if you do a lot of traveling. Being at the right place at the right time has a lot to do with your progress in terms of lucky breaks and opportunities. In 2012, many Gemini will be forced to reevaluate the way they handle their inner life and their health. Many Gemini will also be offered (*or will chose*) the option to relocate and reach a much bigger world in the process; The knowledge found in Astro-Carto-Graphy would be a major contribution to your success (*or your failure*) in one of these new locations. Keep this opportunity in mind and give it a try — it works! http://www.drturi.com/readings/ - Good luck to all of those born in December.

He is wise who understands that the stars are luminaries, created as signs. He who conquers the stars will hold the golden keys to God's mysterious universe.

— Nostradamus —

PERSONAL OBSERVATIONS

www.ingramcontent.com/pod-product-compliance
Lightning Source LLC
Chambersburg PA
CBHW031644170426
43195CB00035B/578